WHAT IS A NUMBER?

12345

67890

What Is a Number?

Mathematical Concepts and

Their Origins **Robert Tubbs**

THE JOHNS HOPKINS UNIVERSITY PRESS   Baltimore

© 2009 The Johns Hopkins University Press
All rights reserved. Published 2009
Printed in the United States of America on acid-free paper
9 8 7 6 5 4 3 2 1

The Johns Hopkins University Press
2715 North Charles Street
Baltimore, Maryland 21218–4363
www.press.jhu.edu

Library of Congress Cataloging-in-Publication Data
Tubbs, Robert, 1954–
What is a number? : mathematical concepts and their origins /
Robert Tubbs.
    p. cm.
Includes bibliographical references and index.
ISBN-13: 978-0-8018-9017-8 (hardcover : acid-free paper)
ISBN-13: 978-0-8018-9018-5 (pbk. : acid-free paper)
ISBN-10: 0-8018-9017-9 (hardcover : acid-free paper)
ISBN-10: 0-8018-9018-7 (pbk. : acid-free paper)
1. Mathematics—Philosophy. 2. Logic, Symbolic and mathematical.
3. Mathematics—History. I. Title.
QA8.4.T83 2008
510.1—dc22   2008013745

A catalog record for this book is available from the British Library.

Special discounts are available for bulk purchases of this book.
For more information, please contact Special Sales at 410-516-6936
or specialsales@press.jhu.edu.

The Johns Hopkins University Press uses environmentally friendly
book materials, including recycled text paper that is composed of at
least 30 percent post-consumer waste, whenever possible. All of our
book papers are acid-free, and our jackets and covers are printed on
paper with recycled content.

For my wife, Vesa

CONTENTS

PREFACE

Nonmathematicians often perceive mathematics as a self-contained, inaccessible body of knowledge that was essentially completed many years, or even many centuries, ago. Mathematics is seen as being isolated from most disciplines, especially from the humanistic endeavors of theology, philosophy, literature, and art. The concerns of mathematicians are imagined to have little in common with those of humanists. But this understanding of mathematics is wrong on two counts: Mathematical objects and goals continue to evolve, and mathematics is not now, and never has been, as separated from the more humanistic disciplines as it might appear.

In fact, at least since the sixth century B.C., theologians, philosophers, writers, poets, and artists have appealed to mathematical ideas and principles to inspire their work and further their arguments. For example the irrationality of the square root of two and the Pythagorean theorem have been used to support theological and philosophical conclusions. At other times, these appeals have been not to precise theorems but to general mathematical concepts, such as the continuum or orthogonality, to provide the intellectual underpinnings for an artistic style or an aesthetic theory.

This book offers an examination of the evolution of a few mathematical concepts—number, geometric truth, infinity, and proof—and of the roles they have played in our continuing attempts to understand the cosmos and our place in it. Using examples from ancient through modern times, this book reveals the central role mathematical notions have played in the history of ideas. Moreover, some of the examples used here illustrate how subtle mathematical relationships, such as the one between a line segment and the points it contains, have challenged both mathematicians and humanists. Through these historical examples, we discover that mathematical ideas are not esoteric ones, divorced from other intellectual or artistic pursuits, but are dynamic ones intrinsic to almost every human endeavor.

The research for this book began during a visit to Williams College in January 2000. I thank both the mathematics faculty at Williams, especially Edward Burger, and the students in my monthlong course for providing me with an intellectually stimulating environment. I also thank my editor Trevor Lipscombe and acquisitions assistant Bronwyn Madeo at the Johns Hopkins University Press and my copy editor Anne R. Gibbons for their expertise and good judgment. My colleague Eric Stade provided invaluable assistance with many of the book's illustrations. I extend my gratitude to Billy Collins for his generous permission to reprint portions of his poem "Questions about Angels." Finally, I thank my wife, Vesa, for her companionship, encouragement, and support. I could not have completed this project without her.

"Lysergic Acid," from Allen Ginsberg, *Collected Poems, 1947–1980* (New York: HarperCollins, 1984). Reprinted by permission of HarperCollins Publishers.

"Tonight I Can Write," from Pablo Neruda, *Selected Poems*, translated by Nathaniel Tarn, Anthony Kerrigan, W. S. Merwin, and Alastair Reid, edited by Nathaniel Tarn (London: Jonathan Cape, 1970). Reprinted by permission of the Random House Group.

"Avatars of the Tortoise" and "The Library of Babel," from Jorge Luis Borges, *Labyrinths: Selected Stories and Other Writings*, translated by James E. Irby (New York: New Directions, 1964). Reprinted by permission of New Directions Publishing Corp. and Pollinger Limited.

"Questions about Angels," from Billy Collins, *Questions about Angels* (Pittsburgh: University of Pittsburgh Press, 1999). Reprinted by permission of the University of Pittsburgh Press and Billy Collins.

WHAT IS A NUMBER?

1

Mysticism, Number, and Geometry: An Introduction to Pythagoreanism

Music has long ... provided the metaphors of choice for those puzzling over questions of cosmic concern. From the ancient Pythagorean "music of the spheres" to the "harmonies of nature" that have guided inquiry through the ages, we have collectively sought the song of nature in the gentle wanderings of celestial bodies and the riotous fulminations of subatomic particles. With the discovery of superstring theory, musical metaphors take on a startling reality, for the theory suggests that the microscopic landscape is suffused with tiny strings whose vibrational patterns orchestrate the evolution of the cosmos.
— *Brian Greene,* The Elegant Universe *(1999)*

In the same year that Columbus discovered the New World and Leonardo sketched the first flying machine, Franchino Gafori opened his influential *Theorica Musice* with a discussion of the musical discoveries of a sixth-century B.C. mathematician and mystic, Pythagoras of Samos.[1] Although Pythagoras is now remembered primarily as a mathematician or early music theorist, he was one of the most important philosophers for the Renaissance, and his ideas have continued to influence Western scientific, metaphysical, and artistic thought. Indeed, had Pythagoras' influence been limited to the Renaissance it is unlikely the twentieth-century writer Arthur Koestler would have written that Pythagoras' "influence on the ideas, and thereby on the destiny, of the human race was probably greater than that of any single man before or after him."[2]

Pythagoras' importance to the history of ideas, and so his impact on "the destiny of the human race," is not a consequence of his discoveries in music or mathematics. Although those discoveries are significant, Pythagoras' influence derives from his philosophical speculations, many of which follow from two metaphysical ideals. The first of these is that the cosmos is not a contingent collection of beings, objects, and bodies but has an underlying mathematical structure; the second is that the fundamental principle, or force, organizing the cosmos is harmony, rather than chaos or coincidence. Taken together, these two beliefs yield alluring equivalencies of truth with order, beauty with harmony, and harmony with mathematical proportion, all of which have guided the development of natural philosophy, theology, literature, art, and mathematics for more than two and a half millennia.

MUSIC AND TRUTH

Heard melodies are sweet, but those unheard
Are sweeter.
— *John Keats, "Ode on a Grecian Urn" (1820)*

The only evidence Pythagoras needed to corroborate his certainty of the essential roles of harmony and mathematics in the workings of the cosmos was his serendipitous discovery of a correspondence between harmonious sounds and mathematical ratios. While there are many variations of this story, the one most widely circulated throughout Europe, before and during the Renaissance, was given by the early fifth-century Roman philosopher Macrobius (395–423). Pythagoras happened to hear the hammerings of a few blacksmiths and noticed that the different blacksmiths' hammers made different sounds: "Thinking that the difference might be ascribed to the strength of the smiths [Pythagoras] requested them to change hammers. Hereupon the difference in tones did not stay with the men but followed the hammers."[3]

Pythagoras then had hammers made of various weights and embarked on an investigation of the relationship between the weights of hammers and the sounds they make when striking an anvil. He discovered that two hammers would produce harmonious sounds only when the weights of the hammers were proportional, that is only when the ratios of their weights equaled a ratio of two whole numbers, such as $2/1$ or $3/4$. Macrobius also described how Pythagoras experimented with large and small bells, with glasses of water filled to various levels, with strings stretched taut by light and heavy rocks, and with hollow pipes of differing lengths. In each case, harmonious sounds were produced when the appropriate quantities, the size of the bells or the weights of the rocks stretching the strings, were in the same proportions as the weights of the blacksmiths' hammers.

This narrative was illustrated by a woodcut in Gafori's text, which attributed the original encounter with the blacksmiths to the biblical figure Tubal-cain. According to this pictorial narrative, as a final act of experimentation Pythagoras turned to a musical instrument called a monochord, which medieval drawings depict as having a movable bridge and a string, or strings, stretched across a wooden sound box. This arrangement allowed the musician, or Pythagoras, to control the pitch of the plucked string by controlling the length of its vibrating segment. Pythagoras measured the lengths of these vibrating strings; he found that harmonious sounds are produced by the same mathematical ratios he had discovered in the blacksmiths' shop.

This may seem to be a rather small discovery, as we now understand the physics of a vibrating string and that a string's pitch is associated with the frequency of its vibration. We also understand that what we perceive as harmony is partly determined by our culture; different tunings or scales have appeared at different times in different societies. So the significance of Pythagoras' discovery is not that he quantified what the musicians of sixth-century B.C. Greece must have already known, at least intuitively. Pythagoras' influence derives from the nature of his reasoning and from his ultimate appeal to beauty or harmony as sources of truth.

As a first example of his speculations, consider what Pythagoras made of his modest musical discovery. Pythagoras could have simply noted the correspondence between mathematical ratios and musical harmonies and accepted that the reason for such a correspondence eluded explanation. (This is what Isaac Newton did in the seventeenth century when he discovered his law for the gravitational attraction between two bodies, which required action at a distance, but offered no explanation for gravity's existence.) Instead, Pythagoras sought a reason for what he had discovered. He concluded that mathematical ratios and musical harmonies are not simply connected; they are equivalent. It is not just that strings vibrating in certain ratios give rise to musical harmonies, but the harmonious sounds are mathematical ratios. Stated in Pythagoras' most extreme terms, mathematical relations not only tell us how to produce harmonious sounds but the sounds themselves are mathematical entities.

This could easily have been the limit of Pythagoras' inferences, and insofar as both music and mathematics are ethereal, almost other-worldly entities, Pythagoras' ideas would have forever remained detached from the material world. But Pythagoras went further: He used the equivalence of musical harmonies and mathematical ratios to connect mathematics with the workings of the universe. This connection depends on a discovery every child has made—if a string is tied to a rock and the rock is whirled overhead it produces a sound. From this observation, Pythagoras concluded that the movements of celestial bodies are associated with sounds, and as these sounds are in the heavens, they must be the purest imaginable musical tones. Reasoning from his belief that musical notes are mathematical entities, Pythagoras concluded that it was not the motion of the planets that made heavenly music but the music that moved the planets.

These sounds are the *music of the spheres*, which the contemporary physicist and mathematician Brian Greene suggests provided the Pythagoreans with a metaphor for understanding celestial motions. And it is reasonable to assume that lacking our modern, scientific language, Pythagoras would give a metaphoric explanation for planetary motion. But Pythagoras was speaking literally; he believed that music choreographed the movements of the sun, moon, planets, and stars. From this, he concluded that an understanding of celestial motions could be obtained through an understanding of their guiding music. Using his earlier conclusion, that musical harmonies are mathematical, Pythagoras argued that celestial motions could be entirely understood through a study of numbers and their ratios.

This is not simply the modern view that mathematical tools and techniques can aid in the analysis of almost any phenomena or that knowledge can be gained through the study of a mathematical model of a physical or social system. Pythagoras inferred from his understanding of musical harmony and celestial motion that every aspect of the physical world is mathematical and can be understood through mathematical principles. For Pythagoras, mathematics did not provide a model for understanding physical phenomena; instead, there is a correspondence between the physical and the mathematical worlds and we can learn about the material world through the study of mathematics. Put slightly differently, mathematics was not just called upon to explain the world; rather, the mathematical world with its objects and relations is one and the same as the physical world.

Even if Pythagoras had ended his metaphysical speculations with his postulation of the existence of the music of the spheres and not gone on to conclude that mathematical principles guide cosmology, he still would have been a significant figure for the Renaissance, and his ideas would have continued to influence poets and artists. As early as the sixth century, the Roman philosopher Boethius (c. 480–c. 526) incorporated the Pythagorean understanding of music into a more expansive scheme. In his *De Musica* Boethius described three different types of music: *musica instrumentalis*, *musica mundana*, and *musica humana*.[4] The first of these is the ordinary music Pythagoras studied following his visit to the blacksmiths' shop; the other two derive from the Pythagorean belief that music resides at the core of all material and human activity. Musica mundana is an extension of the Pythagorean music of the spheres to the entire cosmos; musica humana is the music of our souls, which is essential to our spirituality and existence.

Because musica mundana and musica humana are inaudible, and associated with the heavens and our souls, poets and theologians have repeatedly appealed to these unheard melodies to express the power of their art or belief, for example, in the poem of the English poet John Keats (1795–1821) that introduces this section. A century after Gafori's book appeared, the English poet Philip Sidney (1554–86) invoked the exquisite beauty of musica mundana to defend poetry against Puritan attacks. The Puritans' disagreement was not with poets who attempted to describe divine magnificence or the place of humanity in the cosmos, but by the late sixteenth century, some critics saw English poetry as nothing more than rhyming entertainment in the form of ribald comedies and satirical pieces. The Puritans, reacting to the base appeal of this poetry, embraced Plato's fourth-century B.C. suggestion that poetry be banned from the Republic; it was left to the poets themselves to either defend their art as a whole or explain the difference between salacious rhyming prose and great literature.

Sidney embraced all poetry. In an essay published after his death, *The Defense of Poesy* (1595), now commonly referred to as *An Apology for Poetry*, Sidney equated understanding poetry to hearing the Pythagorean music of the spheres and wrote that those who do not appreciate poetry are dull and are denied the most sublime of all earthly pleasures; poetry connects the human spirit with the heavens. He went even further, saying that if "you cannot hear the planet-like music of poetry; if you have so earth-creeping a mind that it cannot lift itself up to look to the sky of poetry ... [then] when you die, your memory [should] die from the earth for want of an epitaph."[5]

William Shakespeare also appealed to the music of the spheres as evidence for the harmonious working of the cosmos. Toward the end of *The Merchant of Venice*, Lorenzo and his lady, Jessica, are in the garden at night. Lorenzo says to Jessica,

> Look how the floor of heaven
> Is thick inlaid with patines of bright gold.
> There's not the smallest orb which thou behold'st
> But in his motion like an angel sings,
> Still quiring to the young-ey'd cherubins.

In the next three lines Lorenzo speaks of the music in our souls, our musica humana, which we cannot hear because of our mortal nature.

> Such harmony is in immortal souls;
> But whilst this muddy vesture of decay
> Doth grossly close it in, we cannot hear it.[6]

The English poet John Donne (1572–1631), in one of his later poems, "Hymn to God, My God, in My Sickness," wrote of preparing himself for death by preparing his soul's harmony, his musica humana, to match the musica mundana in the heavens:

> Since I am coming to that Holy room,
>   Where, with Thy choir of saints for evermore,
> I shall be made Thy music; as I come
>   I tune the instrument here at the door,
>   And what I must do then, think here before.[7]

Augustine adapted Pythagoras' equivalence of music and number to explain the pleasure obtained from musica instrumentalis. According to the twentieth-century Renaissance scholar S. K. Heninger, "Augustine assumes that there are numbers in the soul, archetypal patterns, and the soul is pleased . . . when sounds reiterate these numbers. A sympathetic vibration is produced resulting in delight to the soul."[8] We feel pleasure from this resonance; our delight emerges from the harmony between the external musica instrumentalis and our elemental musica humana. While this may seem to be an arcane aesthetic theory, especially since modern music theory and psychology have attempted to obtain a rational understanding of music, it is still allowed that our experiences with music are emotional. Indeed, Susanne Langer, in her influential, mid-twentieth-century aesthetic theory, published as *Feeling and Form* (1953), attempted to explain why our most basic responses to music are necessarily emotional and not rational or intellectual. According to Langer, the forms of music match innate structures of our emotional lives. In Langer's words, "Music is a tonal analogue of emotive life."[9] These innate structures are remarkably similar to Augustine's numbers, and this is not a coincidence: these theories are evidence of the lasting influence of Pythagorean ideals.

The Beat poet Allen Ginsberg (1926–97) anticipated that we should perhaps add a fourth type of music to Boethius' list, the subatomic music of modern string theory. In his poem "Lysergic Acid" (1959) Ginsberg sought to hear these harmonies through a drug-induced mystical experience:

> I allen Ginsberg a separate consciousness
> I who want to be God
> I who want to hear the infinite minutest vibration of eternal
> harmony[10]

A quarter century after Ginsberg wrote "Lysergic Acid," physicists affirmed that the properties of the smallest-imaginable objects in the cosmos are determined, in part, by their vibrations. In the current theory, a string cannot exist without its vibrational mode, and since vibrations are the source of music, these different vibrational patterns can be imagined to produce different musical notes. In this most modern of all theories of matter, the string and the music are inseparable; music determines the structure of matter. According to Greene, modern scientists have rediscovered the Pythagorean harmony; the sounds Ginsberg sought to hear by altering his consciousness, string theorists seek to quantify, and hear, through rational, empirical investigation.

THE PHILOSOPHY OF PYTHAGORAS

Discourse not of Pythagorean things without light.
— *Iamblichus, "Protrepticae Orationes ad Philosophiam"*
(3rd century)

Our discussion of the mathematical accomplishments, or philosophical positions, of Pythagoras must begin with a small disclaimer. At the core of this study is an examination of the influence of Pythagorean principles on the history of the idea of truth. A complication in carrying out this analysis is that if Pythagoras wrote anything it has not survived; our knowledge of his ideas comes from the writings of his followers, his presumed influence on other philosophers, and the claims of later commentators. Consequently, there cannot be a clear delineation of Pythagoras' personal philosophy from those of others, and so when we speak of Pythagorean ideas or principles, as we have already done, we do so without intending to ascribe them to Pythagoras himself.

What is known is that Pythagoras established a colony in Elea, Italy, a full two centuries before Plato founded his academy in Athens, and this colony became the center of Pythagoras' religious teachings and then the home for a cult that survived him. Some of these details come to us from a poem in which the Roman poet Ovid (43 B.C.–A.D.17) surveyed the history of the world from its origin out of chaos to the reign of Julius Caesar:

> There was a man here, Samian born, but he
> Had fled from Samos, for he hated tyrants
> And chose, instead, an exile's lot. His thought
> Reached far aloft, to the great gods in Heaven,
> And his imagination looked on visions
> Beyond his mortal sight.[11]

Pythagoras' religion was a mixture of mystical beliefs and concrete, strangely specific, commandments known as *symbola*. This section began with symbolon number 12 from a list of 39 compiled by the third-century Syrian Neoplatonist Iamblichus. A sampling from that list reveals the range of topics Pythagoras is said to have addressed:

no. 3. Sacrifice and worship barefoot.

no. 7. When the winds blow, worship the noise.

no. 8. Cut not fire with a sword.

no. 15. Urin[ate] not, being turned towards the Sun.

no. 22. Wear not a ring.

no. 35. Take not a woman that hath gold, to get children of her.

no. 37. Abstain from beans[12]

Many of these symbola might at first appear to be whimsical; however, they came to be greatly admired. In his highly regarded *Life of Pythagoras* (1706), the classical scholar André Dacier wrote that a symbolon "has an Advantage over a Proverb, as being more concise and figurative, and containing a Moral more delicate and perfect."[13] Some of these symbola can be given straightforward interpretations. For example, in the first century Plutarch took symbolon number 8 to mean not to provoke a man who is already angry.[14] Some symbola are less clear, and consequently have been analyzed by many authors. In the late second or early third century, Diogenes Laertius, the biographer of Greek philosophers, wrote that Aristotle, in a lost book, *On the Pythagoreans*, had given several different interpretations of symbolon number 37 "abstain from beans": "Pythagoras enjoyn'd abstaining from [beans] . . . because they resembl'd . . . the Gates of Hell, as wanting Knees, the Symbols of Mercy and Compassion; . . . or because they are made use of in all Governments, by many Persons, where the Magistrates are chosen by Lots."[15] Plutarch believed that this symbolon spoke to the tradition of using beans as a means of voting: Eating a bean or two would unfairly skew the results.[16] Cicero, however, believed that beans were singled out because "this food is very flatulent, and contrary to that tranquility of mind which a truth-seeking spirit should possess."[17]

Pythagoras' symbola do not appear to have influenced later religious thinkers, but two of his religious beliefs did—the transmigration of souls and monotheism. The first of these, *metempsychosis*, is the belief that every creature is endowed with a soul that following death, returns to earth, possibly within an entirely different species. The specific details of Pythagoras' version of metempsychosis are not known; we do not know which individuals had souls that returned or if souls cycled through both animal and plant life. We also do not know if this was an endless cycle or if there was some ultimate resolution.

Later authors offered their interpretations of the transmigration of souls. A century after Pythagoras, the influential philosopher Empedocles described a reincarnation cycle, "For I have already become a boy and a girl, and a bush and a bird and a fish [corrupt text] from the sea."[18] (Empedocles used metempsychosis to argue in favor of vegetarianism; since souls return in the form of various animals, it would be possible for a father to kill, and eat, his reincarnated son.) Plato also offered a version of metempsychosis, theorizing that every soul has an associated star. "To ensure fair treatment for each," every soul's first incarnation is as a man. If that person lives a good life, the soul returns to its star. Otherwise, until it subdues "all that multitude of riotous and irrational feelings which have clung to it," the soul is reincarnated as a woman, then as "some animal suitable to [its] particular kind of wrongdoing."[19] In *Metamorphoses*, Ovid explained that

> All things are always changing,
> But nothing dies. The spirit comes and goes,
> Is housed wherever it wills, shifts residence
> From beasts to men, from men to beasts, but always
> It keeps on living.[20]

The other aspect of Pythagoras' religion, his monotheism, is more closely related to his conception of mathematics—it is, in large part, a consequence of his understanding of numbers. For Pythagoreans, and all early Greek mathematicians, a number was what we now call a whole number, one of the quantities 1, 2, 3, and so forth. (The early Greeks did not accept negative numbers or have a symbol for zero.) Pythagoras did not even accept 1 as a number; he thought of it as some sort of a primitive element, generating numbers through the operation of addition: $1 + 1 = 2$, $1 + 1 + 1 = 3$. As we will see in chapter 10, this concept of number persisted until the seventeenth century when the geometric continuum was adopted as a model for the collection of numbers (yielding the familiar *number line*).

For the Pythagoreans, just as *one* had meaning beyond being a number so did *two*, *three*, *four*, and each of the other entities *one* generated. The belief that numbers have meanings was still widely held in the Renaissance. In the sixteenth century, the French poet Salluste du Bartas (1544–90) presented an overview of the Renaissance understanding of the Pythagorean conception of numbers. He began with the meaning of *one*:

> the right
> Root of all Number; and of Infinite:
> Loves happinesse, the praise of Harmony,
> Nurc'rie of All, and end of *Polymny*:
> No Number, but more then a Number yet;
> Potentially in all, and all in it.[21]

Salluste du Bartas' association of *one* with "the Infinite" was not based on the simple observation that *one* generated all the counting numbers and that the list of counting numbers is without end, rather it indicated that *one* was considered to be the mystical source of everything; *one* was "Potentially in all" while having "all in it." We examine this use of "the Infinite" in chapter 3 where we label it the metaphysically infinite.

Salluste du Bartas also provided the meanings of *two*, "One's heire apparent / As his first-borne; first Number, and the Parent / Of Female pairs," of *three*, "Th' eldest of odds, . . . / The first that hath beginning, midst, and end," and of *four*:

> a full and perfect summ,
> Whose added parts just unto Tenne doe come;
> Number of God's great Name, Seasons, Complexions,
> Winds, Elements, and Cardinall Perfections.[22]

Since *one* was not a number, *two* was considered to be the first number and the "Parent Of Female pairs." *Three* was considered to be the first odd number and was the number of the male and of harmony. *Four* was not only the number of the square, but it was the number of "God's great Name," because after writing God's Hebrew name, J-E-H-O-V-A-H, and removing the conventionally less significant vowels, God's name consists of the four letters J-H-V-H.

Salluste du Bartas also hinted at a central idea in Pythagorean number theory—that representing a number as a combination of other numbers can reveal additional properties of the number. According to Salluste du Bartas, "The *Ten*, which doth all Numbers' force combine." By this he meant that through the representation $10 = 1 + 2 + 3 + 4$ we discover that ten combines the origin of all numbers, female, male, and divinity or perfection.[23] The Pythagoreans themselves adopted ten as their sacred number and represented it as the five-pointed star that has ten vertices.

FIGURE 1.1. The ten vertices are the five corners of the interior pentagon plus the five points of the star.

Ten also derived a geometric comprehensiveness from the equation $10 = 1 + 2 + 3 + 4$. To understand this it is necessary to understand the Pythagorean, geometric meaning of each number. Geometrically, one was the number of the dimensionless position of a point; two for a segment or line, three for a polygon, or figure in the plane, and four for volume and shape.

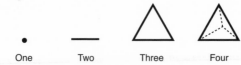

One   Two   Three   Four

FIGURE 1.2. The line segment connects two points; the triangle, three points; and the tetrahedron, four points.

So, ten contained, or manifested, all dimensions and was inclusive of all geometric forms. Ten shared this property with the cosmos and so was assumed to have cosmological significance.

The Pythagoreans knew about the sun, the moon, and the planets Mercury, Venus, Mars, Jupiter, and Saturn. In addition to these seven observable bodies, the Pythagoreans also thought of the earth and the sphere of the stars as heavenly bodies. This yielded a total of nine bodies, one short of the divine ten. The Pythagoreans needed to adjust the observed cosmology to fit the mathematical divinity of ten. In his *Metaphysics* Aristotle wrote that as the Pythagoreans thought the number ten was perfect, "they say that the bodies which move through the heavens are ten, but as the visible bodies are only nine, to meet this they invent a tenth—the 'counter-earth.'"[24]

In the sixth century Simplicius (c. 490–c. 560), who continued to study philosophy in Athens although it had been forbidden in 529, explained why this *counterearth* had never been seen: The Pythagoreans did not believe the earth was at the center of the universe "but that there is a fire in the centre of the universe. And they say that the counterearth, which is an earth, moves around the centre . . . and they say that the earth comes after the counterearth and it, too, moves around the centre. . . . [The counterearth] is not seen by us because the body of the earth always stands in front of us."[25] Thus, for the Pythagoreans, all heavenly bodies, including the earth, rotated around a central fire. This central fire was not considered to be a body; instead it was the driving force for the universe, and it has never been seen because the inhabited portion of the earth, conveniently, always faces the other direction. The counterearth was always on the opposite side of the central fire from the earth.

Pythagoras' universe, with its moving earth, did not survive into the Renaissance, but his number mysticism did. From the first century through the Middle Ages and Renaissance, the significance of the number four was explored through catalogs of the ways it relates to, and possibly guides, the world. Salluste du Bartas alluded to this when he wrote that four is the "Number of God's great Name, Seasons, Complexions, Winds, Elements, and Cardinall Perfections." In the first century A.D. the Greek Theon of Smyrna (c. 70–c. 135) listed various categories that were organized into precisely four possibilities, as if these distinctions were not entirely linguistic. Furthering Pythagorean number mysticism, he found exactly ten categories:

Numbers: 1, 2, 3, 4.
Magnitudes: point, line, surface (i.e., triangle), and volume (i.e., pyramid).
Simple Bodies: fire, air, water, and earth.
Figures of Simple Bodies: pyramid, octahedron, icosahedron, cube.
Living Things: seed, growth in length, in breadth, in thickness.

Societies: man, village, city, nation.
Faculties: reason, knowledge, opinion, sensation.
Parts of the Living Creature: body and the three parts of the soul.
Seasons of the Year: spring, summer, autumn, winter.
Ages: infancy, youth, manhood, old age.[26]

By the sixteenth and seventeenth centuries, the importance of four had been extended to metaphysics (with its four basic principles: *essence, being, power,* and *motion*), to physiology (with the four humors: *phlegm, yellow bile, black bile,* and *blood,* and with the four temperaments: *sanguine, choleric, melancholic,* and *phlegmatic*), and to physics (with the four natural motions: *up, down, forward,* and *circular*). Even in modern psychology there are considered to be four basic personality types: *sensing judges, sensing perceivers, intuitive thinkers,* and *intuitive feelers.*

For the mystic/visionary Romantic poet William Blake (1757–1827), four stood for perfection. The twentieth-century poet Ezra Pound (1885–1972) wrote that four stood for Creation.[27] In his poem "Numbers," Robert Creeley (1926–2005) offered non-Pythagorean descriptions of each of the numbers from zero through nine. Creeley wrote that four "is a square, / or a peaceful circle," an allusion to the ancient problem of trying to convert a circle into a square using entirely geometric methods.[28] (If it were possible to "square the circle," then the circle and square would have the same associated number and be mystically equivalent. The inability of the Greeks, or anyone else, to accomplish this construction enshrouded it in mysticism. Mathematicians demonstrated the impossibility of squaring the circle in the nineteenth century, but this proof required an understanding of number significantly more sophisticated than that of the Greeks and is taken up in chapter 9.)

For now, let's see why the Pythagoreans might have imagined that numbers were geometric, and hence material, objects. There are many ways to represent a whole number. A number can be indicated by marks:

one | two ‖   three ‖| four ‖‖   five ‖‖|

or by pebbles, or dots. Putting aside for the moment the Pythagorean belief that the material and mathematical worlds are one and the same, it is possible to see how representing whole numbers by arrangements of pebbles may have led the Pythagoreans to believe numbers have geometric shapes—four pebbles could be used to illustrate a square and five pebbles could be used to illustrate a pentagon. Although these geometric configurations would be flat, some numbers can be given three-dimensional representations. The most familiar of these are numbers that can be represented as cubes, that is, numbers that represent the number of dots needed to form a geometric cube, such as the eight dots in Figure 1.3.

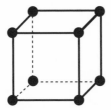

FIGURE 1.3. A cube can be formed using $8 = 2^3$, $27 = 3^3$, or $64 = 4^3$ dots. That these numbers are called *cubes*, just as the numbers $4 = 2^2$, $9 = 3^2$, and $16 = 4^2$ are called *squares* because they can be arranged to make squares, is a holdover from Pythagoras.

From their geometric representations it is easy to imagine that numbers are geometric objects, and since geometric objects are material entities, numbers are as well. This is especially easy to believe for numbers having three-dimensional representations. Because numbers are real, and mathematical reality and physical reality are the same, the Pythagoreans believed matter consists of whole numbers. However, the whole numbers themselves are not material and the points have no size, but numbers give rise to substance because they are associated with material, geometric arrangements.

It is possible to continue this reasoning: If numbers have meaning and are the essential constituents of matter, then every material object has an associated number. Although, in his *Metaphysics*, Aristotle was highly critical of the belief that numbers ascribe properties to material entities, it was generally held to be correct for two thousand years. In his exploration of the relationship between wisdom and mathematical truths, Augustine wrote, "Every material object, however mean, has its numbers." And, according to Augustine, these numbers enable us to judge objects "since we perceive the numbers that are stamped upon them."[29]

As late as the seventeenth century Galileo still felt compelled to address the possibility that numbers have meaning. In Galileo's *Dialogue Concerning the Two Chief World Systems*, which compared the Aristotelian and Copernican cosmologies (which are examined in chapters 4 and 5), Simplicio asked whether or not three is perfect because "there is no passing beyond the three dimensions, length, breadth, and thickness; and that therefore the body, or solid, which has them all, is perfect?"[30] Galileo, through Salviati, replied that "whatever has a beginning, middle, and end may and ought to be called perfect. . . . [But] I feel no compulsion to grant that the number three is a perfect number, nor that it has a faculty of conferring perfection upon its possessors. I do not even understand, let alone believe, that with respect to legs, for example, the number three is more perfect than four or two."[31]

The extension of number properties to objects in the material world is the basis for a now-obscure form of divination known as *geomancy*. The origin of geomancy is unclear, although it has been attributed to Eurytas, an early follower of Pythagoras. A reading proceeds in two steps: the individual seeking information, or guidance, randomly produces an array of stones, or points, then the geomancer gleans the significant patterns from this array and interprets them. This process evolved into a more formal system wherein each meaningful configuration came to consist of four rows, each containing one or two points (there are sixteen such possibilities).

Heinrich Agrippa (1486–1535) illustrated these figures in his three-volume study of occult sciences, *Three Books of Occult Philosophy*, and associated each with an element (earth, air, fire, or water), with one or two of the known heavenly bodies (the moon, sun, Mercury, Venus, Mars, Jupiter, or Saturn), and with one of the twelve astrological signs.[32] Two of these figures are *Puella* and *Rubeus*, below.

FIGURE 1.4. In Agrippa's text, Puella (*left*), was associated with Venus, Taurus, and water, and Rubeus (*right*), with Mars, Scorpio, and fire.

In "The Knight's Tale," Geoffrey Chaucer (c. 1343–1400) appealed to these two geomancy figures to connect the actions on earth with the heavens:

The statue of Mars upon a carte stood
Armed, and looked grym as he were wood,
And over his heed ther shynen two figures
Of sterres, that been cleped in scriptures,
That oon Puella, that oother Rubeus—
This god of armes was arrayed thus.[33]

Although Pythagoras' number mysticism did not survive, his ascribing meaning to each of the counting numbers, 1, 2, 3, and so forth, set them apart from other numerical entities such as ½ or –1 and so hindered the attempts of mathematicians, philosophers, and even theologians to answer the question "what is a number?" But the Pythagoreans themselves discovered a conflict between their assumption regarding the centrality of the counting numbers and one of their most basic geometric truths.

PYTHAGOREAN MATHEMATICS

Greek mathematics is usually identified with geometry. Euclid's third-century B.C. codification of geometric principles in his *Elements* has been widely studied since its translation into Latin in the twelfth century. Yet, Euclid's *Elements* was not just about geometry. Four of its thirteen books were devoted to the study of ratio and proportion, and to the properties of whole numbers. For Pythagoras, geometry and number theory were not unrelated areas. He believed that the only numbers were the whole numbers and that any two geometric magnitudes could be compared using whole numbers or their ratios; so a study of geometric magnitudes required a study of numbers and of ratios and proportions. The Pythagoreans themselves discovered that the conjunction of the beliefs that the only numbers are whole numbers and that all geometric relationships can be expressed using numbers is untenable. Specifically, the Pythagoreans learned that if the whole numbers are the only allowable numbers, then there are geometric lengths that cannot be compared, and conversely, if numbers suffice to compare all geometric lengths then there are numbers other than whole numbers.

What was wrong with the Pythagorean view of mathematics was not so much its understanding of geometric objects and principles, but rather its conception of geometric measurement. This can be illustrated with an example from elementary geometry. Two triangles are *similar* if they have the same shape, in other words if one of them looks like the other one only magnified. Imagine that two triangles are known to have the same shape and that two sides of the first triangle are known, but only one side of the second triangle is known:

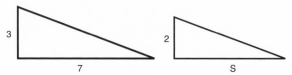

FIGURE 1.5. Two triangles having the same shape, where the length of a side of one of the triangles is not known.

The question is, What can be concluded about the magnitude S? The only way to find S, without simply measuring it with a ruler, is to use results from geometry concerning similar triangles. The Greeks knew that in two similar triangles the ratios of corresponding sides are equal; so using this result for the triangles above yields, in modern notation, $5/7 = 2/3$. To the Pythagoreans, this proportion was the last step in coming to an understanding of S; the proportion told the Pythagoreans that the relationship between the unknown base and 7 is the same as the relationship between 2 and 3. The modern point of view is to think of S as a number and cross-multiply to obtain: $S = 14/3$.

The Pythagoreans would not have accepted the modern measurement of $14/3$ because it is not a whole number. However, the Pythagoreans assumed that it was possible to assign a whole number to any mathematical length, such as the length S. The only catch is that it might be necessary to use a different ruler. To see this idea at work, consider the two triangles in Figure 1.5. The lengths of the sides of these triangles have been measured with a particular ruler, which is taken to be one unit long. If we were to remeasure the sides of these two triangles using a rule whose length is one-third the length of the original ruler then the measurements of these triangles would be

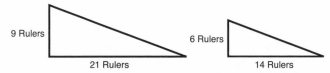

9 Rulers

6 Rulers

21 Rulers

14 Rulers

FIGURE 1.6. If geometric lengths can be measured using fractions, then, by using a different ruler, the lengths can be measured using whole numbers.

The Pythagoreans went beyond this work with similar triangles to make a fundamental assumption: that any two lengths could be comeasured with a single ruler; that is, there will always be a ruler that allows you to simultaneously assign whole-number lengths to any two geometric lengths. This belief is the Pythagorean *commensurability* assumption and is so important to the evolution of the concept of number that it is worth displaying:

> Given any two segments, the ratio of their lengths
> equals a ratio of whole numbers.

This rather technical-sounding statement is an expression of the Pythagorean faith in order and harmony as the organizing principles of both the material and mathematical worlds.

Empirical, or mathematical, evidence has frequently been at odds with aesthetic values; the Pythagoreans themselves discovered the very geometric relationship that undermined their commensurability assumption. And it is possible to describe what the Pythagoreans discovered without an appeal to any sort of mathematical calculation: The measurements of the side and diagonal of a square always violate the commensurability assumption. No matter which ruler is used, the measurement of one of these lengths will always require a less-than-full portion of the ruler. If a ruler precisely measures the length of a side of a square, L, then it will not quite capture the length of a diagonal of a square, D, and if a ruler precisely measures D it will not exactly measure L. The reason for this is the mathematical result uncovered by the Pythagoreans: the ratio of the length of a side to the length of the diagonal, L/D, can never equal the ratio of two whole numbers. In our language, this is the statement that L/D is not a rational number; this ratio of geometric lengths is something the Greeks did not accept as a number.

The spirit of this book is not just to describe conflicts between mathematical and aesthetic, theological, scientific, or artistic ideas, but where possible to explain the mathematical reasons for these conflicts. The explanation of how the Pythagoreans established the incommensurability of the side and diagonal of any square removes the mystery from the result—it is simply a mathematical deduction requiring a short calculation based on what Johannes Kepler (1571–1630) would later call "the golden theorem of Pythagoras on the squares of the sides in a right-angled triangle" and what is known to us as the Pythagorean theorem.[34]

The Pythagorean theorem establishes a simple relationship between the lengths of the three sides of any triangle, provided one of its angles equals 90 degrees. If A and B denote the lengths of the two shorter sides, and C denotes the length of the side opposite the 90 degree angle, then $A^2 + B^2 = C^2$.

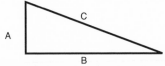

FIGURE 1.7. The Pythagorean theorem states that in a right triangle with legs of lengths A and B, and with a hypotenuse of length C, the relationship $A^2 + B^2 = C^2$ always holds.

The Pythagorean theorem is one of the first results in geometry whose truth is not immediately self-evident. The reason the Pythagorean theorem is true is because we can construct a convincing argument to support it. This argument is based on a simple geometric construction. The first step is to take four copies of the above triangle and one copy of a square whose sides are all C units long and rearrange the four triangles and the C by C square:

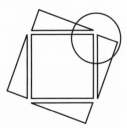

FIGURE 1.8. The three angles within the circle add up to 180 degrees, as they are the three angles in the original triangle. So if the four triangles are glued to the central square the five figures will form a larger square (because the sides of this larger figure will be straight lines).

The deduction of the Pythagorean theorem from the square in Figure 1.8 then depends on the following observation: The total area contained in the original four triangles and C by C square is the same as the area of the larger square they comprise. If these two areas are calculated and equated, a bit of algebra leads to the formula known as the Pythagorean theorem: $A^2 + B^2 = C^2$. What will be important later in this book is not this demonstration but the knowledge that it depended on the angles in the original triangle adding up to 180 degrees.

The Pythagoreans now had two ways to examine the relationship between the square's side and diagonal, the Pythagorean theorem and the commensurability assumption. Combining these two relationships it is possible to obtain a single equation

$$2 = (a/b)^2$$

where a and b are whole numbers. But a bit of experimentation shows that no matter which whole numbers a and b are tried, $(a/b)^2$ will never be precisely equal to 2. Looking at one hundred or even one thousand examples does not show that there are not some exotically large whole numbers a and b with $2 = (a/b)^2$, but the Pythagoreans themselves proved mathematically that no such whole numbers can exist. In our notation, the Pythagoreans discovered that the square root of two does not equal any fraction; in modern terminology, the square root of two is an *irrational* number.

This irrationality of the square root of 2 told the Pythagoreans that it is impossible to capture the relationship between the side and diagonal of a square through the use of whole numbers and ratio and proportion. The side and diagonal of a square are an example of what was not supposed to exist, incommensurable geometric magnitudes.

GEOMETRY AND NUMBER

While the discovery of incommensurable lengths undermined the Pythagorean conception of mathematics, it did not weaken the Pythagoreans' faith in the equivalence of physical and mathematical reality. Thus, this discovery did not precipitate a crisis, except for Hippasus (born c. 500 B.C.), who is said to have been drowned at sea after uncovering it. But the existence of incommensurable lengths did force the Greeks to acknowledge a conflict between their assumptions that the only numbers are whole numbers and that all geometric lengths can be understood through whole numbers and their ratios. Rather than simply expand their concept of number, and accept that some quantities are outside the realm of the harmonious ratios of whole numbers, the Greeks chose to pursue geometry without numbers, thus separating the study of mathematics into two subjects: arithmetic (the study of properties of numbers) and geometry. Having abandoned the original Pythagorean principle, *all is number*, the Greeks turned to geometry for both their understanding of the material world and their aesthetic theory.

2

The Elgin Marbles and Plato's Geometric Chemistry

There are natural Causes of Beauty. Beauty is a Harmony of Objects, begetting Pleasure by the Eye. There are two Causes of Beauty—natural and customary. Natural is from Geometry.... Always the true test is natural or geometrical Beauty.

Geometrical Figures are naturally more beautiful than other irregular [figures]; in this all consent as to a Law of Nature.

— Sir Christopher Wren, "Appendix: Of Architecture," Parentalia *(1750)*

Chapter 1 examined the influence of the presumed equivalence of the material and mathematical worlds on attempts to understand the cosmos, and in particular on the belief that the universe is guided by harmoniously interacting mathematical structures. But there is another conception of reality that has greatly influenced our attempts to understand the universe that also relies on mathematical perfection. This view of reality is the belief that beyond the world of our experience, there is another, more real, possibly more important world. The material world is contingent and mutable; absolute beauty, or truth, resides only in the other inaccessible realm. In his important book *The Great Chain of Being* (1936), A. O. Lovejoy labeled this belief *otherworldliness* and gave it the following definition: "[It is] the belief that both the genuinely 'real' and the truly good are radically antithetic in their essential characteristics to anything to be found in man's natural life."[1]

This chapter explains how otherworldliness and Pythagorean principles blended to yield one of the most mathematically appealing of all attempts to comprehend the tangible world. The geometric conception of matter given by Plato in the fourth century B.C. Plato's theory is not just a historical curiosity; it relies on two ideas the twentieth-century physicist Werner Heisenberg pointed out still influence the course of scientific investigations: "the conviction that matter consists of minute indivisible units" and "the belief in the purposely directive power of mathematical structures."[2]

By way of contrast, Pythagoras' belief in numbers as the basic constituents of matter does not manifest either of Heisenberg's principles; Pythagoras' numbers were not material objects governed by mathematical principles but geometric configurations whose properties were determined through number mysticism.

GEOMETRIC BEAUTY

In 1817 Keats attended an exhibit of marble sculptures that Thomas Bruce, the Seventh Earl of Elgin, had arranged to bring to England from the Parthenon. Among these sculptures was *Three Goddesses*.

PLATE 2.1. *Three Goddesses* (Hestia, Dione, and Aphrodite) (frieze from the east pediment of the Parthenon). Phidias (c. 490–430 B.C.). British Museum, London. Photo: Scala / Art Resource, New York.

Upon viewing these statues, Keats was seized by the realization that material beauty can provide a portal to the hidden truths of a more significant world beyond our experience. He published his reaction to the exhibit a few days later in his poem "On Seeing the Elgin Marbles."

My spirit is too weak—mortality
Weighs heavily on me like unwilling sleep,
And each imagin'd pinnacle and steep
Of godlike hardship, tells me I must die
Like a sick eagle looking at the sky.

Yet 'tis a gentle luxury to weep
That I have not the cloudy winds to keep
Fresh for the opening of the morning's eye.
Such dim-conceived glories of the brain
Bring round the heart an undescribable feud;
So do these wonders a most dizzy pain,
That mingles Grecian grandeur with the rude
Wasting of old Time—with a billowy main—
A sun—a shadow of a magnitude.

In a letter to his friend Benjamin Bailey, Keats explained what he believed to be the source of these statues' power: "What the imagination seizes as Beauty must be truth—whether it existed before or not—for I have the same Idea of all our Passions as of Love they are all in their sublime, creative of essential Beauty."[3] Keats concluded a later poem, "Ode on a Grecian Urn," with a clear expression of this aesthetic theory in a famous equivalence:

When old age shall this generation waste,
Thou shalt remain, in midst of other woe
Than ours, a friend to man, to whom thou say'st,
"Beauty is truth, truth beauty,—that is all
Ye know on earth, and all ye need to know."[4]

Pythagoras began with the axiom

reality is mathematical

and with an unshakable belief in harmony as the organizing principle of the cosmos. Combining Pythagoras' axiom with Keats' equivalence

beauty is truth, truth beauty

yields the Pythagorean principle

truth is mathematically harmonious.

This is the aesthetic value adopted by Greek artists and architects, and espoused by Wren (1632–1723) in the quotation that introduces this chapter.

Wren was the architect whose ideas remade the profile of London following the destructive fire of 1666. In his architectural theory, Wren distinguished natural beauty from customary beauty; natural beauty arises from geometric harmony, and customary beauty from culture and experience. Geometric beauty is more important than customary beauty precisely because mathematical truths are more fundamental than coincidental, material ones. For example, Wren believed the Gothic designs of the great cathedrals of Europe were chaotic and confused, and the use of nooks and crannies conformed to custom and not to natural beauty (for example, in the Milan cathedral, late fourteenth to early sixteenth century, or in the Notre Dame cathedral in Paris, twelfth century). Wren's Saint Paul's Cathedral has clearer lines; its geometric form is more evident and manifests the more fundamental beauty Wren sought. Wren, however, was not an otherworldly thinker; Wren simply accepted what mathematicians, cosmologists, artists, and scientists alike have continued to accept—the profundity of mathematically harmonious truths.

There is another aspect of Wren's architectural theory that is, perhaps, more psychological than philosophical, the notion that specific geometric forms are particularly attractive, and so profound. The Pythagoreans also held some geometric forms to be more beautiful than others; to appreciate the mathematical aesthetics underlying these forms it is necessary to return briefly to Pythagorean mathematics.

PYTHAGOREAN GEOMETRY

Pythagorean, and more generally Greek, geometry had two components: the deduction of new geometric truths from known ones and the explicit construction of geometric figures and relationships. For the Greeks, a geometric construction was more appealing, and so more acceptable, if it could be performed using only a straightedge and a compass, that is by either drawing a line segment or an arc. Measurement was not allowed, the Pythagorean discovery of incommensurable lengths had made it suspect.

In *The Secret of the Universe* (1596), Kepler referred to the Pythagorean theorem as "the golden theorem of Pythagoras." Kepler continued with the assertion that there is another "treasury of geometry, on the line divided in the extreme and mean proportion."[5] The second geometric result Kepler so admired illustrates the sort of balance and harmony the Pythagoreans sought to establish through geometric methods. The precise result may not at first seem to be especially attractive—the Pythagoreans developed a geometric procedure for dividing any segment into two pieces whose lengths satisfy a special relationship. Specifically, given the segment from a point A to a point B, the Pythagoreans showed how to find a point P on the segment so that the lengths of the segments AB, AP, and PB satisfy a special relationship: the proportion $AP/AB = PB/AP$. The modern point of view is to take the location of P as an unknown, view the proportion $AP/AB = PB/AP$ as an equation, and use the rules of algebra to solve the equation. (The Pythagoreans did not have algebraic techniques; Arab mathematicians invented algebra a millennium later.)

Part of the Pythagoreans' interest in the proportion $AP/AB = PB/AP$ is its connection with their sacred star, below. In this star each line that crosses one of the longer segments divides it into its extreme and mean ratio. Specifically, the point P divides the side AB into its extreme and mean proportions, as does the point Q. Thus, according to the Pythagoreans, the star is beautiful, in part, because it manifests these mathematical proportions.

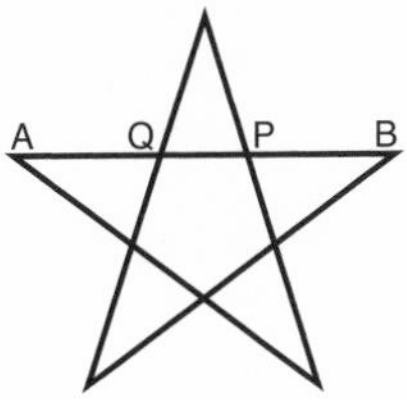

FIGURE 2.1 The Pythagorean sacred star.

The extreme and mean proportionals are also precisely those that give rise to the so-called golden rectangle. The ratio PB/AP is the same as the ratio of the width to length of any golden rectangle; so a golden rectangle can be formed by taking any segment, finding the point P giving the extreme and mean ratio, and bending the segment at P to form two sides of a rectangle. Golden rectangles, which are purported to be the most visually pleasing of all rectangular forms, can be discerned in both architecture and art. In architecture, golden rectangles can be superimposed on photographs of the Parthenon and the United Nations building in New York. Representations of golden rectangles are not quite as evident in art, but several authors have claimed to have found them, for example in Seurat's *Circus Sideshow* (below) and Mondrian's *Place de la Concorde* (1943). For a convincing refutation of these claims see Mario Livio's book *The Golden Ratio: The Story of Phi, the World's Most Astonishing Number.*[6]

PLATE 2.2. *Circus Sideshow*, 1887–88. Georges Seurat (1859–91). Oil on canvas, 39¼ × 59 in. (99.7 × 149.9 cm). Bequest of Stephen C. Clark, 1960 (61.101.17). Photo: Bruce Schwarz. The Metropolitan Museum of Art, New York. Image © The Metropolitan Museum of Art / Art Resource, New York.

PLATO'S THEORY OF CREATION

God pour'd the Waters on the fruitfull Ground

In sundry figures; some in fashion round,

Som square, som cross, som long, som lozenge-wise,

Some triangles, som large, som lesser size.

— *Salluste du Bartas, "The Third Day of the First Weeke,"* His Devine Weekes and Workes *(16th century)*

Plato's Pythagoreanism is expressed almost exclusively in the *Timaeus*, which he is said to have written after traveling to Italy to meet with two of the remaining Pythagoreans (the astronomer Timaeus and the mathematician Archytas). In the *Timaeus*, Plato discusses an impressive array of topics, including cosmology, a theory of matter, human psychology, and the lost culture of Atlantis. Plato's understanding of matter was tied to its origin, and as his theory must accommodate the principles of his creation myth, it is natural to begin there.

Before the *Timaeus*, the creation of the universe was based on an organic model involving either birth or growth; part of Plato's originality lies in his assertion that there was a single creator for all things. The

Creator, or Maker, first established a structure for the heavens and then produced the substance of the material world. In performing these acts, the Creator employed mathematical principles: Number mysticism and harmony determined the form of the heavens; aesthetics and geometry, the structure of matter.

To form the cosmos, the Creator positioned circles of *world soul* around the earth at distances determined by the squares and the cubes of the so-called Pythagorean numbers one, two, and three (so using the numbers $1 = 1^2 = 1^3$, $4 = 2^2$, $9 = 3^2$, $8 = 2^3$, and $27 = 3^3$). The Creator filled the rings between these circles with more strips of world soul at distances determined by mathematically harmonious values. Plato described these distances in terms of mathematical ratios, writing that the Creator fills the intervals between the original circles by inserting "two mean terms in each interval, one exceeding one extreme and being exceeded by the other by the same fraction of the extremes, the other exceeding and being exceeded by the same numerical amount."[7]

Having established a cosmic framework for the universe, the Creator turned to the construction of its material. To understand the Creator's role here it is necessary to recall that Plato distinguished between two worlds—the world of perfect forms and the imperfect world of experience. The otherworldly realm of forms, of mathematical objects and truths, is the world of *being*; the objects in the material world are weak approximations of their perfect, *real* forms. The Creator modeled the construction of matter on mathematically beautiful forms from the world of *being*, attaching triangles together to produce harmonious geometric structures for the four elements: earth, air, fire, and water.

Plato did not explain the origin of the Creator, but he did say that the Creator's power derived from his *goodness*. Because the Creator is *good*, he combined chaotic substance in mathematical proportions to produce the universe and the basic components of matter. This Creator shares attributes with the Western, Christian God. This is not so much a coincidence as an indication of Plato's influence. By the fourth century, Augustine had already noted this connection; he praised Plato, along with the third-century Neoplatonists, for their conception of a creator of all things. In *The City of God*, a description of the world's history from Genesis to Last Judgment, Augustine wrote that because of God's "immutability" and "simplicity," the Platonists "realized that God is the Creator from whom all other beings derive, while he is himself uncreated and underivative." Augustine equated Plato's Creator and the Christian God, and went on to explain how a pagan could have obtained such knowledge without access to Scripture by quoting the Bible, Romans 1:19–20: "What can be known of God has been revealed among them. God in fact has revealed it to them. For his invisible realities, . . . have been made visible to the intelligence through his created works, as well as his eternal power and divinity."[8]

In a letter to Saint Jerome, Augustine even took up the analogy of God creating the world as a musician makes music, so mirroring Plato's placement of the strips of world soul: "If a man who is skilled in composing a song knows what lengths to assign to what tones, so that the melody flows and progresses with beauty by a succession of slow and rapid tones, how much more true is it that God permits no periods of time in the birth and death of His creatures—periods which are like the words and syllables in the measure of this temporal life—to proceed either more quickly or more slowly than the recognized and well-defined law of rhythm requires, in this wonderful song of succeeding events."[9]

In Plato's theory, the Creator's ability to form order from chaos emerged from his *goodness*. When transferred to Christian thought, this goodness was often associated with benevolence and with the preference for order over disorder. Late in the sixteenth century, in "An Hymn in Honour of Love" (1596), the poet Edmund Spenser (1552–99) appealed to Plato's description of the Creator's role to describe God's creation of order from elemental chaos. Before God's intervention,

> The earth, the air, the water, and the fire,
> Then 'gan to range themselves in huge array,
> And with contrary forces to conspire
> Each against other, by all means they may,
> Threat'ning their own confusion and decay:
> Air hated earth, and water hated fire,
> Till Love relented their rebellious ire.[10]

For both Plato and Spenser, the world was created from chaos. In the *Timaeus* the chaotic world of the *receptacle of becoming* did not belong to our world; it was apart from the material world of experience. In Spenser's poem the chaotic world was the material world before the intervention of the Christian God. Unlike Plato's Pythagorean world, the one Spenser's Creator produced from the primordial material of the receptacle of becoming did not necessarily have to be harmonious; an omnipotent God could just as well have created a hell on earth as a sympathetic cosmos. As Spenser explained in the next stanza of his poem, God chose to make the cosmos harmonious, and he did so through an application of his benevolence, which Spenser called *love*. For a discussion of parallels between Plato's Creator and Renaissance poetic theories, and views of the poet, see this chapter's postscript.

THE NATURE OF SUBSTANCE

Plato's description of the building blocks for matter accommodated three Greek theories: that everything is essentially mathematical, that all matter consists of some primary substance or substances, and that all objects are made up of indivisible *atoms*. To appreciate how seamlessly Plato integrated these ideas into a unified whole, this section begins with a brief overview of primary-substance and atomic theories.

The seventh-century B.C. natural philosopher Thales proposed that water was the essential component, the *arche* or first principle, of all matter. To the modern-day mind this sounds absurd. But in light of the ancients' limited scientific knowledge the choice of water is not that farfetched. Thales understood that if there were only a single fundamental substance it must be malleable and capable of moving from place to place, and water satisfies these requirements.

The primary-substance theory was continued by Anaximander (c. 610–546 B.C.), for whom it was air, and by Heraclitus (c. 500 B.C.), for whom it was fire. These two theories were more nuanced than Thales'. Anaximander understood that if there were only one substance, it would fill up all of space in a uniform manner and no object could be discriminated from any other; he introduced the idea that matter emerges from a homogeneous existence through the attraction and repulsion of opposites. Heraclitus introduced another concept, which became part of Plato's theory, that of transformation:

> As all things change to fire,
> and fire exhausted
> falls back into things.[11]

A generation before Plato, Empedocles had augmented this model by taking earth, air, fire, and water as the fundamental elements. For Empedocles, nature composed the material of the world as a painter composes a picture:

> As when painters adorn votive offerings,
> men well-learned in their craft because of cunning,
> and so when they take in their hands many-coloured pigments,
> mixing them in harmony, some more, others less,
> from them they prepare forms resembling all things,
> making trees and men and women
> and beasts and birds and water-nourished fish
> and long-lived gods, first in their prerogatives.[12]

There were two forces at work in Empedocles' theory, *love* and *strife*; these two forces were as primitive as are the elements, earth, air, fire, water (this theory should be thought of as having six components, the two forces plus four elements). Love combines elements, and so produces harmony; strife moves elements apart, and so produces chaos and disorder.

The third idea Plato blended into his theory of matter was borrowed from two of his contemporaries, Leucippus and Democritus. Leucippus, followed by Democritus, proposed that all matter consisted of atoms. For them, an atom was an indivisible piece of matter perpetually in motion within empty space. Leucippus' atoms were unlimited in shape and number. Democritus contributed to this atomic theory by developing its epistemological implications. According to Democritus, "In reality we know nothing; for truth is in the depths."[13]

Democritus believed that the only properties of matter we can perceive are ephemeral ones we impose on short-lived arrangements of atoms: "By convention sweet and by convention bitter, by convention hot, by convention cold, by convention colour; but in reality atoms and void."[14] Foreshadowing Allen Ginsberg's desire to understand reality by hearing the harmonies of the "minutest vibration[s]," Democritus wrote that there are two forms of knowledge, one "genuine" (the one Ginsberg sought) and the other "bastard": "To the bastard form belong all of these, sight, hearing, smell, taste, touch, but the other is genuine and separate from this. When the bastard form can no longer see anything smaller or hear or smell or taste or perceive by touch, but to a finer degree ..."[15] Thus, according to Democritus, we can have genuine knowledge of an object, or of matter, only through an understanding of its constituent atoms.

There is no sense of mathematical perfection in the Leucippus/ Democritus atomic theory. Their atoms had an unlimited number of forms and they did not necessarily move along mathematically harmonious paths. To a sometimes-Pythagorean like Plato an atomic theory offering chaotic motion and chaotic shapes as the basis of all matter was no more acceptable than Anaximander's appeal to the chaotic interaction of opposites. Moreover, Plato did not quite embrace the Leucippus/ Democritus early manifestation of the reductionist principle, that knowledge comes from explaining events, or objects, by relating them to a variety of forms or laws. Instead, Plato suggested that the form of an atom influences its properties, but the true nature of material was not to be found through an investigation of its constituent atoms, because these are only approximations of perfect, otherworldly forms. Nonetheless, Plato adopted part of the Leucippus/Democritus atomic theory; he postulated that there are four, not necessarily indivisible, *atoms*, one for each of the fundamental elements, earth, air, fire, and water, but their otherworldly forms were understood entirely through mathematical principles.

Plato began the development of his theory of matter with a metaphysical question: How do we know matter exists? Given Plato's tendency to be an otherworldly idealist, his response was surprisingly empirical: We know matter exists because we can see it and feel it. Of these experiences, the first requires that there be light, the second that there be a component of tangibility. Before Robert Boyle's experiments with luminous bacteria and fungi in the seventeenth century, fire was the only known source of light. Fire was thought to produce candlelight, sunlight, and lightning, so Plato explained that light, that is, fire, must be a fundamental element. Tangibility was more problematic as it has multiple manifestations, sand, stone, trees, and cats, for example. Rather than explore what these material objects have in common, Plato only sought to find the element that distinguished them, and everything terrestrial, from the heavens. Plato took Empedocles' multipurpose earth as this element.

Next, instead of appealing to our experiences with the atmosphere, wind, and sea to conclude the cosmos contains Empedocles' other two elements, air and water, Plato took a more mystical and geometric approach. Plato claimed that "it is not possible to combine two things properly without a third to act as a bond to hold them together," but he did not appeal to a metaphysical force, such as Empedocles' love. Instead, Plato wrote that "the best bond is one that effects the closest unity between itself and the terms it is combining; and this is best done by a continued geometrical proportion."[16]

In our discussion of the extreme and mean proportional, we found a point P dividing a line segment AB into two segments, whose lengths satisfy a certain proportion. This problem has an analogue for numbers: Given two whole numbers a and b, find a whole number x, between a and b, so that $a/x = x/b$. (If we let r equal this ratio, that is, if we let $r = a/x = x/b$, a bit of arithmetic reveals that $x = (1/r) \times a$ and $b = (1/r)^2 \times a$; in this situation the numbers a, x, and b are said to be in *continued geometric proportion* since they are part of a geometric sequence.) For arbitrary whole numbers a and b, it is not possible to find an appropriate whole number x, but it is possible to find such an x when a and b are squares (i.e., if a and b each equals a whole number squared). In this case $x = \sqrt{ab}$.

Plato apparently knew this mathematical result, because he appealed to the Pythagorean idea that square numbers are flat two-dimensional objects, while cubes are solid three-dimensional objects, to justify requiring two bonding agents between elements, instead of one: "If then the body of the universe were required to be a plane surface with no depth, one middle term would have been enough to connect it with the other terms, but in fact it needs to be a solid, and solids always have two connecting middle terms."[17] Translated into the language of ratio and proportion, Plato used the mathematical result that if a and b are whole numbers, which are cubes (solids), then it is always possible to find whole numbers x and y, between a and b, so that

$$a/x = x/y = y/b.$$

In this case $x = \sqrt[3]{a^2 b}$ and $y = \sqrt[3]{ab^2}$. (Again, if we let r equal this ratio, $r = a/x = x/y = y/b$, then the numbers a, x, y, and b will be in continued geometric proportion: $x = (1/r) \times a$, $y = (1/r)^2 \times a$, and $b = (1/r)^3 \times a$.)

Thus mathematics led Plato to the conclusion that there must be four elements, or at least provided him with some justification for using four elements. According to Plato, the Creator "placed water and air between fire and earth, and made them so far as possible proportional to one another, so that air is to water as water is to earth."[18]

Although Plato imitated the Leucippus/Democritus theory of indivisible atoms, contrary to that theory, Plato's elements were not immutable. In Plato's chemistry, fire consumes earth, or at least some manifestations of earth such as wood, and water extinguishes fire. This is an elaboration of Heraclitus' view:

Air dies giving birth
to fire. Fire dies
giving birth to air. Water
thus, is born of dying
earth, and earth of water.[19]

Plato viewed these processes as transformations and gave an explicit cycle for the four fundamental elements:

$$\text{water} \rightarrow \text{earth} \rightarrow \text{air} \rightarrow \text{fire} \rightarrow \text{air} \rightarrow \text{water}$$

Plato explained this cycle as: Water solidifies to become earth; earth dissolves and evaporates to become air; through combustion air becomes fire, which when extinguished and allowed to condense becomes air; finally, air contracts and condenses to "cloud and mist," which, when compacted, becomes water.[20]

This left Plato with the following challenge—to uncover forms of the four primary elements that both mirror their properties and accommodate, or explain, the transformation cycle. The same mathematical result upon which Plato based his theory ultimately undermined it.

THE GEOMETRY OF PLATO'S MATTER

Because each of the four elements in Plato's theory transforms into another, he did not speak of the constituents of matter as "being a thing" but of "having a quality," and the (mathematical) forms of these elements determine their qualities.[21] Nonetheless, Plato established a geometric shape for each of them and attempted to base the transformation cycle on geometric principles. As earth, air, fire, and water are the basis of all material existence the aesthetics Plato inherited from the Pythagoreans led him to believe that these elements' true forms must be mathematically beautiful. Rather than take number mysticism as a basis for that beauty, Plato chose harmony as manifested through geometric symmetry.

There are an unlimited number of highly symmetric solids; they range from the familiar cube to one of those mirrored balls, spinning from the ceiling of a discotheque. Among this multitude of possible surfaces Plato sought the most beautiful four. Without acknowledging it, Plato made several restrictive assumptions concerning the shape of these fundamental surfaces:

> every face (or side) must be the same regular polygon (in a *regular* polygon all sides have the same length and every angle has the same measurement);
> adjacent faces must all meet in the same angle; and
> every corner of the surface must look like every other corner of the surface.

These restrictions mean that if the solid is viewed from one side, or corner, it will look the same as when viewed from any other side, or corner.

There are only five such highly symmetric, geometric objects, a mathematical result known to Plato (see Figure 2.2). Three of these five solids have faces that are equilateral triangles (the tetrahedron, octahedron, and icosahedron); one has faces that are squares (the cube); one has faces that are regular pentagons (the dodecahedron).

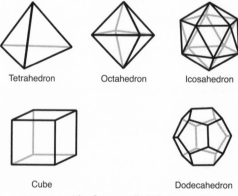

Tetrahedron   Octahedron   Icosahedron

Cube   Dodecahedron

FIGURE 2.2. The five so-called Platonic solids.

The mathematical result that there are only five so-called Platonic solids is fairly easy to establish, but all we really need to know, for later reference, is that the Greek proof that there are only five Platonic solids depends on the same geometric relationship as the Pythagorean theorem (see chapter 1):

The sum of the angles in any triangle equals 180 degrees.

By considering the qualities of these solids, such as their stability or sharpness, Plato associated them with earth, air, fire, and water. As regards their stability Plato wrote: "Let us assign the cube to earth; for it is the most immobile of the four bodies and the most retentive of shape; ... similarly we assign the least mobile of the other figures to water, the most mobile to fire, and the intermediate to air." He then reaffirmed this association by appealing to each solid's sharpness: "we assign ... the sharpest to fire, the next sharpest to air, and the least sharp to water."[22] Thus, Plato assigned the cube to earth, the tetrahedron to air, the octahedron to fire, and the icosahedron to water. The dodecahedron became the cosmos.

In this association the cube, tetrahedron, and so forth are not thought of as solid three-dimensional objects, as if they were carved out of some substance, but as hollow surfaces. The twelve-sided dodecahedron was assumed to represent (or be) the cosmos with its twelve signs of the zodiac. Plato explained that "the [Creator] used [the dodecahedron] for embroidering the constellations on the whole heaven."[23]

PLATO'S GEOMETRIC CHEMISTRY

Chapter 1 describes how the discovery of a mathematical relationship, or a deduced mathematical result, can undermine an aesthetically appealing conception of either the mathematical or material world. The Pythagorean discovery that $\sqrt{2}$ cannot be written as any fraction, or its geometric consequence that there are incommensurable geometric lengths, should have told the Pythagoreans that it was not possible to maintain the equivalence of mathematics and reality and still take harmony as the central organizing principle of each. Instead, they took a narrower view of mathematics; one based on the primacy of geometry and geometric methods. Plato understood this; even so, there is a mathematical conflict embedded in Plato's theory of matter.

This conflict does not involve a subtle assumption about the nature of geometric measurement but is a consequence of the axioms of Greek geometry.

Plato's transformation cycle relied on decomposition and rearrangement: The geometric surface corresponding to an element in the cycle decomposes into polygons and these polygons rearrange into the geometric shape of the next element in the cycle. This works for those steps in the cycle where one triangular-faced solid transforms into another, for example for the air to fire, fire to air, or air to water transformations. Since the number of triangles needed to form each of these elements is different, this yields a primitive form of stoichiometry. In the transformation of air into fire, imagine having a large quantity of air, hence many octahedrons. Each octahedron has eight sides, each of which consists of an equilateral triangle. A single element of air decomposes into eight equilateral triangles, any four of which can combine to form the solid representing fire—the four-sided tetrahedron. So in the cycle each air transforms into two fires. Symbolically: *Air = 2 Fire*. Plato also employed his geometric chemistry to explain what happens "when water is broken up by fire or again by air."[24] The twenty equilateral triangles from the decomposed water-icosahedron recombine to form one four-sided fire-tetrahedron and two eight-sided air-octahedrons. Symbolically: *Water = Fire + 2 Air*.

But Plato needed to explain how the square faces of earth's cube could rearrange to become the triangular faces of air's octahedron. Plato's insight was that each face of the cube could be imagined to consist not simply of a square but of two triangles, obtained when the square is cut along a diagonal. If we adopt this point of view, when an earth-cube decomposes into polygons it decomposes not into eight squares but into sixteen triangles, and these triangles are then supposed to recombine to form the triangular faces of air.

It is here that Plato's aesthetically pleasing transformation cycle breaks down. The same mathematical result that ensures the rarity, and so significance, of the symmetric solids representing earth, air, fire, and water, prohibits this earth-to-air transformation. Specifically, because *the sum of the angles in any triangle equals 180 degrees* it is impossible to recombine the triangles from earth to obtain the equilateral triangles that constitute air. (This is because each earth-triangle has two 45-degree angles and one 90-degree angle while each air-triangle has three 60-degree angles.)

The failure of his geometric mechanism to account for his entire transformation cycle did not convince Plato to admit something might be wrong with his Pythagorean theory of matter. Instead, Plato offered an ad hoc explanation for the earth-to-air step of his transformation cycle—after earth decomposes into triangles, they remain in limbo waiting to re-form into earth-cubes.[25]

THE QUALITIES OF MATTER

Air is moist and warm and, although opposed to water, the cold to the warm, nevertheless has a common bond of moisture. Moreover, fire, being hot and dry, spurns the moisture of air, but yet adheres to it because of the warmth in both.
— *Macrobius,* Commentary on the Dream of Scipio *(c. 400)*

Plato's student Aristotle shared neither his teacher's Pythagorean tendencies nor his otherworldliness. In his *Physics*, Aristotle attempted to explain the relationships between the four elements through their empirical qualities rather than through their hidden mathematical structures.[26] In this approach, each of the four elements had a characteristic temperature, hot or cold, and a characteristic feel, moist or dry. Fire is hot and dry and water is cold and moist. It is less evident, but inarguable, that earth is cold and dry and air is hot and moist. (During the late Middle Ages and Renaissance, the four elements were commonly arranged around a circle or a square, with adjacent elements sharing a property, and opposite qualities, and consequently opposing elements, on opposite sides of the figure.) The proportional relationships between two elements were then determined by whether there was an intermediate element sharing a quality with each. For example, air shares moistness with water and water shares coldness with earth, supporting Plato's claim, from the *Timaeus*, that "air is to water as water is to earth."[27]

Macrobius, in the same text in which he described Pythagoras' discovery of mathematical ratios within musical harmonies, explained the physical consequences of the relations assigned to the elements:

"And so it happens that each one of the elements appears to embrace the two elements bordering on each side of it by single qualities: water binds earth to itself by coldness, and air by moisture; air is allied to water by its moisture, and to fire by warmth; fire mingles with air because of its heat, and with earth because of its dryness; earth is compatible with fire because of its dryness, and with water because of its coldness."[28] Macrobius went on to appeal to the mystical geometric properties of four and to Plato's appeal to mathematical proportions: "These different bonds would have no tenacity, however, if there were only two elements; if there were three the union would be but a weak one; but as there are four elements the bonds are unbreakable, since the two extremes are held together by two means."[29]

Another way to interpret the relationships among the elements is to think of the qualities as being more basic than the elements. In this theory, the qualities combine to form the elements; earth is the only element that is both cold and dry and so is viewed as being the combination of coldness and dryness. Similarly, water is the combination of coldness and moistness; air, of moistness and heat, and fire, of heat and dryness.

This alternate point of view was fully developed in the Renaissance, and in his epic poem *Paradise Lost* (1667), John Milton (1608–74) called upon the opposition of qualities in his description of hell. When Satan arrives at the gates of hell,

> in sudden view appear
> The secrets of the hoary deep; a dark
> Illimitable ocean, without bound,
> Without dimension, where length, breadth, and highth,
> And time, and place, are lost; where eldest Night
> And Chaos, ancestors of Nature, hold
> Eternal anarchy, amidst the noise
> Of endless wars, and by confusion stand:
> For hot, cold, moist, and dry, four champions fierce,
> Strive here for mastery, and to battle bring
> Their embryon atoms.[30]

Milton's understanding of hell was a further development of Spenser's idea that God's love enabled, or caused, him to produce order from chaos. Not only are the souls in hell unable to experience God's love, they are forced to eternally suffer chaos (the absence of harmony).

THE MODERN ATOM

It is fair to ask: Which point of view is correct? Does the structure of matter determine its properties or do universal qualities come together to, in effect, produce matter? The obvious answer, taking Heisenberg's claims into account, is that Plato was essentially correct, but not in the details. The less obvious answer is that Aristotle, and later Renaissance thinkers, were also correct. The brief history of the atom illustrates this.

Owing mostly to the power of Aristotle's natural philosophy, the atomic theory for matter was not widely accepted either in the Middle Ages or Renaissance. However, as observation, theology, and Aristotle's physical theories became more and more divergent, other ideas were occasionally entertained. Chapter 10 describes the fourteenth-century attempts of some Scholastics to revive atomism in defense of theological principles, but these Scholastics were in a minority. The reintroduction of atomic theory into mainstream intellectual discourse is usually attributed to John Dalton, in the early nineteenth century, but it was Pierre Gassendi (1592–1655) who revived atomic theory by making it more acceptable to the theologians of his time.

Before Gassendi, atomism was associated with atheism and paganism because of its pre-Christian origin. Moreover, the acceptance of atoms both contradicts what was the generally accepted conception of infinity and leads to seemingly irresolvable paradoxes. Gassendi did not overcome the Aristotelian objections to atoms, but he did make atoms somewhat theologically acceptable by claiming that they are solids created by God and that they are moved through divine influence.

Almost two centuries later, in 1803, Dalton introduced an atomic theory manifesting both the Platonic and Aristotelian ideas. Dalton's atoms were indivisible, mathematically perfect spheres with measurable masses and, more importantly for Dalton, had qualities that would now be called chemical properties. The chemical property of an atom was not determined by its shape but by an invisible quality, its mass. The atoms of a particular element all had the same mass, and atoms of different elements had different masses.

Not until early in the twentieth century was the atom viewed as rings of electrons revolving around a positively charged central particle, and then as rings of electrons revolving around a nucleus consisting of both positively charged and neutral particles. In this model, the properties of an atom were determined by how it combines the *qualities* of the electron, proton, and neutron, and how the qualities of the electron, in particular, contribute to the properties of the atom was determined by the geometry of the electron rings around the nucleus.

Two developments led to a model significantly more difficult to visualize than this solar system one. The first was the realization that the atomic world is not quite as deterministic as had been thought. This knowledge came from the development of quantum mechanics. In this theory, electrons do not stay in predetermined orbits; their position can only be given probabilistically. Any particular electron could, theoretically, be anywhere in the universe at any given moment; the atom just tends to look like the early twentieth-century model. The second realization was that things that were thought to be indivisible are not. This began with the splitting of the atom, but that was only the first step, the next was the divisibility of an atom's constituent parts.

By the middle of the twentieth century, the atom consisted of electrons, protons, and neutrons. Each of these particles was considered to be indivisible. But by the late twentieth century, these particles were no longer thought to be the most basic. Many more particles had been discovered; there were three families each consisting of four different particles. The electron remained, but the proton and neutron were each seen to consist of three *quarks*: The proton was made up of two up-quarks and one down-quark, the neutron of one up-quark and two down-quarks. Everything in our terrestrial world is made of electrons, up-quarks, and down-quarks; the other particles, neutrinos, muons, and the like, exist only in the heavens in the solar wind and cosmic rays.

Now, even these particles are no longer imagined to be the basic components of the cosmos. That role has been turned over to strings, unimaginably short segments of matter, for lack of a better concept. All strings are of the same stuff, yet they have different properties. The property of a string is determined by qualities, mass and energy, and these qualities are determined by geometry and music. How twisted a string is determines its mass; how rapidly it vibrates determines its energy level. Geometry and music produce the properties of matter from entities so small they defy three-dimensional conceptualization.

POSTSCRIPT: PLATO'S MAKER AND POETIC THEORY

By the time Sidney equated the ability to hear the music of the spheres with the ability to rise above the muck of our daily existence and appreciate the beauty, and reality, of poetry, the role of the poet had been elevated to that of a Platonic Creator. "The Greeks called him [*poieten*], which name hath, as the most excellent, gone through other languages. It cometh of this word . . . which is 'to make'; wherein I know not whether by luck or wisdom we Englishmen have met with the Greeks in calling him a maker."[31]

In late sixteenth-century poetic theory, the poet-as-maker took the raw material of our existence and formed it into understandable substance. Sidney, contrasting the poet with the astronomer, geometer, natural philosopher, grammarian, and others, concluded: "Only the poet, disdaining to be tied to any . . . subjection, lifted up with the vigor of his own invention, doth grow, in effect, into another nature, in making things either better than nature bringeth foorth, or, quite anew, forms such as never were in nature."[32] This view of the poet-as-maker was widely held, and an entire poetic theory emerged from this interpretation of the role of the poet. Ben Jonson expressed this concisely: "[Poetry is] the queen of arts, which had her original from heaven, received thence from the Hebrews and had in prime estimation with the Greeks, transmitted to the Latins and all nations that professed civility."[33]

In her study of seventeenth-century poetry, *The Divine Science*, the modern literary scholar Leah Jonas explained that Jonson's statement captures all of the central points of that century's poetic theory: "(1) the heavenly origins of poetry, (2) the priority of poetry among the arts and sciences, (3) the coincidence of poetry and civilization."[34]

It is only fair to note that Italian poetic theory was already well developed by the end of the sixteenth century and it offered a less otherworldly view of the poet. In an influential seven-volume description of poetic theory, *Poetics*, the sixteenth-century Italian literary theorist Julius Caesar Scaliger provided a more mundane definition of poetry and the poet: "The poetical art is a science, that is, it is a habit of production in accordance with those laws which underlie that symmetrical fashioning known as poetry. So it has three elements—the material [subject], the form [type], and the execution [style]."[35]

In Scaliger's theory the poet is a maker not in the sense of creating new matter or a new reality but in presenting a new vision of reality through verse. A poem must be written "in accordance with those laws that underlie that symmetrical fashioning." The style of the prose determines whether the text is a poem; the excellence of a poem is determined by its subject matter. Scaliger went so far as to provide a ranking of the types of poetry, in decreasing order of significance, from hymns to songs "sung in praise of brave men" to epics, tragedies and comedies, satires, and pastorals.[36]

The English theorists were more generous in what they took to be poetry but still separated mere verse from poetry. Sidney wrote, "[Verse is] but an ornament and no cause to poetry, since there have been many most excellent poets that never versified, and now swarm many versifiers that need never answer to the name poets."[37]

There were political reasons not to equate rhyming verse with poetry: Sidney could only deflect the Puritan attacks by redefining poetry and excluding the ribald and salacious public verse from consideration. Having taken the term "poetry" to mean something different than was in common usage, Sidney had to differentiate between a poem (or poet) and that which is not a poem (or poet). Instead of delineating the differences, Sidney offered three type of poets: The poet who *makes* in the sense of Plato's Creator; the poet who is a seer or visionary and presents new ways of understanding the material world; and the poet who is a teacher who philosophizes. Other writers, whatever they are, are not poets, and among poets, those who dip into the poetic analogue of Plato's receptacle of becoming and create beauty are the ones who provide us access to the truths held by the music of the spheres.

3

An Introduction to Infinity

There is a concept which corrupts and upsets all others. I refer not to Evil, whose limited realm is that of ethics; I refer to the infinite. I once longed to compile its mobile history. The numerous Hydra ... would lend convenient horror to its portico. . . . Five or seven years of metaphysical, theological and mathematical apprenticeship would allow me (perhaps) to plan decorously such a book. It is useless to add that life forbids me that hope or even that adverb.

— Jorge Luis Borges, "Avatars of the Tortoise" (1932)

The twentieth-century Argentinean writer Jorge Luis Borges (1899–1986) began his essay "Avatars of the Tortoise" with the above quotation, acknowledging the complexity of the concept of the infinite. Underlying Borges' introductory paragraph was his realization that infinity is not a monolithic idea. Almost every intellectual, spiritual, or artistic enterprise has, at some time in its history, appealed to the infinite, and within each discipline the infinite has been used at different times, in different ways. The infinite has been invoked by philosophers to describe the true nature of a reality forever outside our direct experience, and by theologians to contrast the divine from the earthly, and so place God beyond human comprehension. For poets, the infinite has represented mystical insight and truth that cannot be gained through rational analysis or scientific theories. Even mathematicians have appealed to the infinite to both reject and defend certain mathematical processes and conclusions. Indeed, the ambivalence of mathematicians toward the acceptance of infinite processes or collections is at the heart of this book.

The purpose of this chapter is to complete in a few pages a portion of the task Borges imagined would take him "five or seven years" just to plan—to delineate one use of the infinite from others in order to establish a foundation for understanding the evolution of its mathematical meaning. And although Borges was right in asserting that it is not possible to offer a definition for the infinite covering every usage, it is possible to distinguish between different types of appeals to the infinite and so appreciate its influence on mathematics, philosophy, religion, science, and art. A first step toward accomplishing this is to isolate three, mostly distinct, categories of the infinite: the qualitative, the quantitative, and the poetic.

POETIC INFINITY

For, though the Lord of all be infinite,
Is his wrath also? Be it, man is not so,
But mortal doom'd. How can he exercise
Wrath without end on man whom death must end?

. . . .

     Will he draw out,
For angers sake, finite to infinite
In punish'd man, to satisfy his rigour
Satisfied never?
— *Milton,* Paradise Lost *(1667)*

Milton employed infinite (or infinity) in two ways—as an attribute of God and as a measure of God's power. In the first instance, infinite expresses a quality, and in the second, a quantity. These two senses of the infinite may appear to be related in that Milton could have been referring to God's size or extent when he wrote, "the Lord of all be infinite," but he was doing something else. Milton was using infinity in an entirely qualitative, metaphysical manner. Milton's metaphysically infinite God is infinite in the sense of being a unified, indivisible reality or being, something transcendent, complete, and independent of experience.

Milton's two uses of infinite must be distinguished from what we have already labeled poetic infinity. When Vincent van Gogh (1853–90) wrote, "a child in the cradle . . . has the infinite in its eyes," he was not using "infinite" in either of the ways we ascribed to Milton.[1] Rather, van Gogh's infinite was meant to evoke innocence, clarity of vision, and mystical understanding. Poetic infinity can sometimes be taken to be a metaphysical infinity, as in the first four lines of William Blake's (1757–1827) poem "Auguries of Innocence" (c. 1803):

> To see a World in a grain of sand,
> And a Heaven in a wild flower,
> Hold Infinity in the palm of your hand,
> And Eternity in an hour.[2]

But Blake's infinity was more than metaphysical. Here infinity is used in a manner that includes both the metaphysically infinite and the mathematically infinite, but it also refers to knowledge and mystical insight. When you "Hold Infinity in the palm of your hand" you have a complete, intuitive understanding of the cosmos and your place in it.

Half a century later, Alfred Tennyson (1809–92), without directly using the term, sought in the infinite not mystical insight but a metaphysical understanding of a Christian God and man's place in the universe:

> Flower in the crannied wall,
> I pluck you out of the crannies,
> I hold you here, root and all, in my hand—
> Little flower—but if I could understand
> What you are, root and all, and all in all,
> I should know what God and man is.[3]

Poetic infinity can also be used to evoke powerful emotions. Our last example of poetic infinity is one of its most effective uses; it is from Pablo Neruda's (1904–73) poem "Tonight I Can Write" (1924). The poem begins, "Tonight I can write the saddest lines," and then Neruda explains that he has lost a woman he once loved, who also once loved him. In the last stanza of the poem, Neruda tells us that his former lover will be another's.

> Another's. She will be another's. Like my kisses before.
> Her voice. Her bright body. Her infinite eyes.[4]

With the phrase "Her infinite eyes," Neruda reveals to the reader how he felt both about her and the eternity he faces knowing they were once together and never will be again.

The remainder of this chapter examines the evolution of the concept of the metaphysically infinite and the earliest theory of mathematical infinitude. This separation of the metaphysical infinite from the mathematical is slightly problematic, as metaphysical infinity has properties that reflect the mathematical, and the two are sometimes confused or used interchangeably. For example, in the thirteenth century Thomas Aquinas (1225–74) described God as being infinite in both senses. To Aquinas, God was metaphysically infinite because he was self-sufficient and perfect, but he was also mathematically infinite because to be finite is to be less than can be imagined, and so less than perfect. These two notions of infinity also were combined in Anselm of Canterbury's eleventh-century definition of God—God is that than which no greater can be conceived. (From this definition Anselm deduced the existence of God, because if God did not exist then the thought that God exists represents a conception of a greater entity.) Despite this tradition of blurring the distinction between metaphysical infinity and mathematical infinity, understanding their differences is necessary to appreciate the role of theological discussions in the evolution of mathematical thought.

A SHORT HISTORY OF METAPHYSICAL INFINITY

The concept of the metaphysically infinite has been invoked to address different philosophical and religious difficulties in Western thought for at least twenty-five centuries. In Hellenistic Greece the metaphysically infinite emerged in attempts to answer: What is the nature of matter and/or reality? In medieval Europe, it played a significant role in answering: What is God? In post-Renaissance philosophy, it has periodically appeared in attempts to answer: How do we know anything? The first era ended when Aristotle untangled mathematical infinity from metaphysical infinity; the second started with the post-Neoplatonist reconciliation of Aristotle's natural philosophy and theological ideals; and the third began with seventeenth-century attempts to integrate the Copernicus-Galileo cosmology with Christian faith.

The Pythagorean universe was governed not just by the harmonious interaction of mathematical structures but also by finite, and orderly, whole numbers and their ratios. However, there was another sixth-century B.C. conception of reality, the philosopher Anaximander's view that everything, all matter and space, emerged from, and consists of, a primordial chaotic material. The fundamental material behind the order and structure of the world was both the chaos of nonexistence and the fundamental element of all matter. Where the Pythagoreans embraced the simplicity of mathematical beauty and harmony, Anaximander invoked the pejorative *to apeiron*.

Like Pythagoras' whole numbers, Anaximander's substance existed in physical form—it was the material from which all other material was made. Anaximander did not offer a mechanism to explain how unorganized chaos transformed itself into flowers, rocks, birds, and oceans, but the Pythagoreans did not offer such a mechanism either. In Anaximander's theory the construction of the material world somehow involved the interaction of opposites. Anaximander described what makes up all things, not why it is so or how it happens.

Anaximander's metaphysically infinite substance shares features with Milton's God—both were unbounded, unknowable sources of everything, and both were more real than the material world of human experience. But Anaximander's substance was not endowed with the positive attributes of Milton's God. More sophisticated appeals to metaphysical infinity were given by Parmenides of Elea (born c. 515 B.C.) and then by Melissus of Samos (born early fifth century B.C.). Parmenides was a Pythagorean who rejected the idea that the world consists of a system of structures within space. In the view of the earlier Pythagoreans, reality emerged from the harmonious interaction of these finite, orderly mathematical structures within a limitless, unbounded void. The void itself was indefinable; it was unordered and without shape or meaning. Parmenides did not question the centrality of mathematics in the structure of the cosmos; indeed it was Parmenides' concern with order that moved him to challenge the earlier Pythagorean model. What Parmenides could not accept was the description of reality as something existing within a void. To Parmenides, this reference to a void allowed for the possibility that chaos (in the form of nothingness) could be essential to the workings of the external world, conflicting with the Pythagorean assumption that order was the unifying principle of nature. To remedy this defect, Parmenides proposed that reality did not exist in a void; instead it must be a whole having nothing outside of itself.

Parmenides' avoidance of emptiness also influenced his conception of the fundamental nature of reality—he concluded that reality is a homogeneous, static entity. To understand how Parmenides reached his conclusion we must adopt a slightly novel view of how motion occurs. Instead of viewing a change in position as a movement through space, think of it as a passage from one state to another. In this view, change is a transition from an existing situation to one that does not yet exist; a condition that does not exist is a physical void waiting to be filled. Thus, change is movement from existence into nothingness, and since this nothingness cannot exist, motion is impossible. From this conclusion Parmenides argued that all states must simultaneously coexist everywhere. Reality must be everywhere the same. Similarly, Parmenides argued that reality must be eternal: It always has existed and always will exist because a beginning requires movement from nothingness and an ending requires movement into nothingness.

Of course, Parmenides could not deny that the observable world is forever in flux, and that change is the rule rather than the exception. This forced him to postulate a radical distinction between reality (the world as it is) and appearance (the world as we see it). Parmenides concluded that everything we perceive is illusionary. Reality is a complete, eternal, homogenous whole that exists beyond our finitude—reality is a metaphysically infinite entity. Parmenides called this reality *what-is*.

Parmenides offered his description of *what-is* in poetic form. Only 154 lines of his poem *On Nature* remain, but this remnant describes a mystical journey in which Parmenides meets a benevolent goddess. The poem begins with Parmenides' journey:

> The mares that carry me as far as my heart ever aspires sped me on, when they had brought and set me on the far-famed road of the god, which bears the man who knows over all cities. . . . And the goddess greeted me kindly, and took my right hand in hers, and addressed me with these words: "Young man, you who come to my house in the company of immortal charioteers with the mares which bear you, greetings. . . . It is proper that you should learn all things, both

the unshaken heart of the well-rounded truth, and the opinions of mortals, in which there is no true reliance."[5]

Parmenides learns from the goddess that *what-is* is a static, unmoving entity. Depending on how one translates the term *ateleston*, *what-is* is either perfect, bounded, or balanced—perhaps all three.

Parmenides' Pythagorean inclination to equate mathematical and physical reality led him, through the goddess, to give *what-is* a geometric representation: "For it must not be any larger or smaller here than there. For (1) neither is there what-is-not, which might prevent it reaching the same distance; (2) nor is there any way that what-is could be more than what-is here and less there, since it is all immune to plundering: for equal to itself on all sides, it has equal being within its limits."[6] According to this passage, *what-is* is a solid entity that cannot be different in one direction from another. Although *what-is* has sides, it extends the same distance in any direction, presumably from a center, and so is the most mathematically perfect of all forms—a sphere. This description of *what-is* as a sphere might lead us to visualize it as a bounded, three-dimensional object existing in some ambient space, but this is the wrong image. Parmenides' point was that *what-is* cannot be visualized. Anything we visualize, or even imagine in our mind's eye, is merely appearance. *What-is* can only be known through mystical insight.

Half a century later, Melissus attempted to give a clearer description of *what-is*. He accepted Parmenides' radical appearance/reality distinction; his only significant departure from Parmenides' view concerned the spatial attributes of metaphysical infinity: "It has no [spatial] beginning or end, but is infinite. For if it had come to be it would have a [spatial] beginning (for it would have begun the process of coming-to-be at some time) and end (for it would have ended the process of coming-to-be at some time). But since it neither began nor ended [the process], and always was and always will be, it has no [spatial] beginning or end."[7] This quote from Melissus is significant: He takes mathematical (quantitative) infinitude as a property of a metaphysical infinity. This is one of the earliest examples of the blurring of the distinction between the qualitatively and quantitatively infinite.

So, in a relatively brief period, from the early sixth century to the middle of the fifth century B.C., the idea of metaphysical infinity evolved from being Anaximander's substance, to Parmenides' unified, homogenous reality beyond our reach, then to Melissus' measure of reality's size.

THE WHOLE IS GREATER THAN THE PART

Infinity turns out to be the opposite of what people say it is. It is not "that which has nothing beyond itself" that is infinite, but "that which always has something beyond itself."

— *Aristotle,* Physics *(4th century B.C.)*

Aristotle (384–322 B.C.) has been called the first philosopher of the infinite, and his ideas about the infinite were shaped, in part, by his desire to clearly delineate mathematics from natural philosophy. Aristotle did not equate the mathematical and material worlds, as had the Pythagoreans; he believed, instead, that mathematical objects were abstractions based on the material world. Indeed, this separation of the material and mathematical worlds is no more evident in any of Aristotle's theories than in his theory of infinity.

Instead of examining the concept of infinity solely in the realm of pure mathematics, Aristotle sought to answer a more straightforward question: Is anything in nature, in the material world of space and time, infinite? In book 3 of his *Physics*, Aristotle dismissed Anaximander's assertion that infinity is a substance through an appeal to one of the most fundamental, and ultimately incorrect, assumptions about the nature of both material and mathematical objects: The whole is greater than the part. Aristotle's argument begins: Any substance must have parts (or portions) and each of these parts must be of the same substance. If infinity is a substance then a portion of it is also infinity, and so by taking a portion of infinity one appears to obtain a greater and a lesser infinity. But there is only one size of infinity, so this conclusion conflicts with the self-evident truth that the whole is greater than a part.[8]

Aristotle also rejected Parmenides' and Melissus' conception of the metaphysically infinite. For each of them, the metaphysically infinite was not only static, it was a whole that has no part outside it. Aristotle thought this understanding of the infinite was exactly backward.

For Aristotle, the infinite is not that which is whole but that which can never be a whole. From this, Aristotle concluded that the metaphysically infinite cannot exist and infinite may only be applied quantitatively; infinitude might be the measure of something, either a magnitude or a collection, but it cannot be one of its qualities.

MODERN METAPHYSICAL INFINITY

This infinite and immobile space which is so certainly discerned in the nature of things will seem . . . to be something not merely real but divine.
— Henry More, "Enchiridion metaphysicum" (1671)

The belief in a transcendent, metaphysically infinite reality did not completely disappear with Aristotle's pronouncements. The second-century philosopher Plotinus was one of the first to challenge Aristotle's rejection of metaphysical infinity. Plotinus resurrected the otherworldliness of Parmenides and Plato, reestablishing the possible existence of a metaphysical infinity. For Plotinus, and this will sound familiar, there was a transcendent realm beyond our experience and the world of our senses. It was a self-sufficient, perfect, omnipotent entity. At times Plotinus referred to this entity as something underlying our reality, associating it with Parmenides' *what-is*; at other times, Plotinus idealized the entity as the One, associating it with Plato's realm of ideals. What is most significant is that Plotinus not only resurrected the metaphysically infinite but also endowed it with creative, positive attributes, freeing it from the unknowable chaotic darkness of early Greek thought and allowing for its later incorporation into a description of the Christian God. Plotinus wrote, "The One is perfect because it seeks nothing, and possesses nothing, and has need of nothing; and being perfect, it overflows, and thus its superabundance produces an Other."[9] Plotinus then explains why the One produced the multitude of existing things: "Whenever anything reaches its own perfection, we see that it cannot endure to remain in itself, but generates and produces some other thing. Not only beings having the power of choice, but also those which are by nature incapable of choice, and even inanimate things, send forth as much of themselves as they can: thus fire emits heat and snow cold."[10] In the third century Augustine imported Plotinus' conception of metaphysical infinitude into Christian theology, and by the time Milton wrote *Paradise Lost* the metaphysical infinitude of the Christian God was unquestioned.

And the metaphysically infinite had other manifestations, both in cosmology and in philosophy. Early in the seventeenth century, Galileo discovered unseen stars and, he thought, planets. Space was suddenly vast, incomprehensible. One of the first English poets and philosophers to consider the consequences of these astronomical discoveries was Henry More (1614–87). More was one of a small group of influential Platonists at Cambridge University in the middle of the seventeenth century. As Marjorie Hope Nicolson explained in her book *Mountain Gloom and Mountain Glory*, "More transferred to space some twenty attributes formerly associated with God." The excerpt from More that heads this section continues with, "The divine names and titles which precisely harmonize with it . . . are these which severally belong to Metaphysically Primal Being." More then listed the twenty attributes of space, which included simplicity, incorruptibility, and incomprehensibility.[11] More was not equating space with his Christian God; it is simply that both are metaphysical infinitudes. The connection between the two emerged from More's belief that in our contemplation of the metaphysically infinite space, we are contemplating God.

Later in the seventeenth century, Joseph Addison (1672–1719) made this same connection. Addison wrote of the pleasures we obtain from our encounters with the vast or unbounded: "Our Imagination loves to be filled with an Object, or to grasp at any thing that is too big for its Capacity. We are flung into a pleasing Astonishment at such unbounded Views, and feel a delightful Stillness and Amazement in the Soul at the Apprehension of them."[12] The awe we feel when we view limitless terrestrial vistas or an overwhelming sequence of mountains upon mountains inspires us to understand the enormity of God's Creation. This, to Addison, was instinctual: God created us so that our realization of the vastness of terrestrial and heavenly space would lead us to a contemplation of the unlimited God. Without the overtly Christian component, this theme was taken up by the Romantic poets of the eighteenth century.

Also in the eighteenth century, Immanuel Kant (1724–1804) based his entire metaphysics on the distinction between appearance and reality. Echoing Parmenides, Kant proposed that we gain knowledge about the metaphysically infinite reality when we receive information. Although we are quantitatively finite beings in an overwhelmingly metaphysically infinite world, we obtain knowledge about the world by letting it impinge upon our senses, thus receiving information that we interpret through our preexisting framework.

In Kant's view, we gain increasingly accurate knowledge of the metaphysically infinite real world through a mathematically infinite process. When we receive information from (for lack of a better expression) the real world, owing to our finite nature, the information we receive must be partial and limited. We cannot take in the metaphysically infinite whole all at once. This observation led Kant to address how we can obtain any knowledge, if all we receive are small, imperfect bits of data. The key to Kant's model is our awareness that what we are receiving is finite and conditioned by the context of our observation. But knowing this context means that we possess additional information, so we know more than the information we received. This sets up an infinite regression, because this larger quantity of information itself is conditioned by some context, which means we have even more information or clearer knowledge. This is an endless process that we can never complete.

As an example, consider how Kant, in his *Critique of Judgment* (1790), explained how it is that we can gauge the size of something: "The magnitude of the measure has to be assumed as a known quantity, if, to form an estimate of this, we must again have recourse to numbers involving another standard for their unit."[13] In other words, Kant believed that whenever we conceive of the size of something, we must already have in mind a particular unit of measurement. That unit itself can only be understood in terms of another unit of measurement (probably of a smaller magnitude) that we already understand. Thus, according to Kant, we can only understand any particular measurement through recourse to an endless number of smaller and smaller units of measurement, and each unit is comparable with the one below it in the hierarchy: "Since in the estimate of magnitude we have to take into account not merely the multiplicity (number of units) but also the magnitude of the unit (the measure), and since the magnitude of this unit in turn always requires something else as its measure and as the standard of its comparison, and so on, we see that the computation of the magnitude of phenomena is, in all cases, utterly incapable of affording us any absolute concept of a magnitude, and can, instead, only afford one that is always based on comparison."[14] Kant claimed that all measurements are commensurable in the informal sense that they are all comparable, but the smallest common unit that could be used to measure any two things can never be achieved. This is because to comprehend the smallest of all possible units, we would have to first understand a yet smaller (or somehow more fundamental) unit of measurement and so on. This seeming infinite regression of comparisons is precisely what separates the finite from the infinite. For Kant the sublime is that which in comparison everything else is small. It is not only that the infinite (sublime) is something that cannot be compared with anything else; the infinite is that which requires no comparison. Our perception of it inspires the same awe as Addison's vistas.

The question for us is whether Kant was discussing a mathematical infinity or a metaphysical infinity. Kant did not think an absolutely great quantity, or object, could exist in nature. Indeed, he said there is nothing in nature that is so large that it cannot be viewed as infinitely small, when compared with something else. This is consistent with his infinite chain of magnitudes. The infinite for Kant was the metaphysically infinite, something complete and beyond comparison, something beyond measurement and so beyond our experience.

Kant's theories for how we come to know anything, because in his view there was an insurmountable obstacle between what reality is and how we experience it, have continued to influence modern philosophical thought. Even those who disagree with Kant's basic appearance/reality distinction have addressed his ideas (even if only to reject them). We will not follow this historical development, which runs through Hegel (1770–1831), Husserl (1859–1938), Bergson (1859–1941), and Wittgenstein (1889–1951); of these, Bergson's philosophy, discussed below, is the most relevant to modern mathematics because it has components of both Kant and Plato.

AN INTRODUCTION TO QUANTITATIVE INFINITY

The universe (which others call the Library) is composed of an indefinite and perhaps infinite number of hexagonal galleries, with vast air shafts between, surrounded by very low railings. From any of the hexagons one can see, interminably, the upper and lower floors.

— Jorge Luis Borges, "The Library of Babel" (1941)

Now that we have an appreciation of the persistence of the metaphysically infinite in philosophy and theology, we return to Aristotle's conception of the mathematically infinite. But before examining Aristotle's subtle, and later controversial, conclusions about infinite quantities, it is helpful to examine one of his less subtle, but also ultimately controversial, conclusions about infinitude. The most natural candidate to be an infinite entity is the entirety of the existing universe. "Existing" is used here to modify universe to avoid the issue of an endlessly evolving cosmos.

As Aristotle's conception of infinitude excluded the possibility that an existing entity, or collection of entities, could be infinite, because an existing object, or collection, is complete and thus does not have any part outside itself, the Aristotelian universe must be finite. This conclusion was based entirely on Aristotle's definition of infinite, but it also follows from the basic principles of his physics.

Not everyone in the ancient or medieval worlds agreed that the entire universe was a finite, completed entity, and this is perhaps the most easily challenged of all of Aristotle's conclusions about infinity. As early as the fifth century B.C., the Greek mathematician Archytas, who was one of the two Pythagoreans Plato is said to have visited before writing the *Timaeus*, had questioned whether a finite universe was possible. Archytas reasoned as follows: If space is not infinite, then the universe must be finite and if the universe is finite then it is bounded. Archytas then asked what would happen if someone, standing at the boundary of the universe, were to reach out with his hand. According to Archytas, one of two things could happen: Either the person's hand would go beyond the universe, and thus the universe could not be finite, or something would stop the person's hand, and that something must be the surface containing the universe. But surfaces have two sides, so there must be something beyond the boundary of the universe.

Archytas' argument appeals to our intuition that a surface divides space into two parts, so if the universe were contained within a surface it would be natural to ask what lay beyond the surface. To Aristotle, this question has no meaning. His solution, to what we perceive as a possible objection to having a finite, bounded universe, was to maintain that if something is limited, it is not necessary that it be limited in relation to something else.[15] This is a very modern mathematical idea, but it was never taken as a thorough refutation of Archytas' proof.

Lucretius (first century B.C.) appealed directly to common sense to refute Aristotle's refutation of Archytas' argument, "It is a matter of observation that one thing is limited by another."[16] This position was implicit in his argument against a bounded universe: "Suppose for a moment that the whole of space were bounded and that someone made his way to its uttermost boundary and threw a flying dart.... Whether there is some obstacle lying on the boundary line that prevents the dart from going farther on its course or whether it flies on beyond, it cannot in fact have started from the boundary." Lucretius' argument relies on our intuition that the universe is the same everywhere, and that the laws of physics hold uniformly throughout the cosmos. The belief that the geometry of the universe is uniform is precisely the one overturned in the nineteenth and twentieth centuries with the discovery of non-Euclidean geometries and the realization that infinite and bounded are not necessarily mutually exclusive attributes. (This is discussed in chapter 8.)

In his story "The Library of Babel" Borges indicated that reasoning that the geometry of the space we have experienced must necessarily be the geometry of all of space has led the Library's inhabitants to conclude that it was endless: "Those who judge [the Library] to be limited postulate that in remote places the corridors and stairways and hexagons can conceivably come to an end—which is absurd."[17] The story's narrator rejects, as absurd, the idea that the universe (Library) could be finite, because every hexagonal room that has been entered has been identical to the others before it, and there has always been a next hexagon. The geometry of the Library must be everywhere the same.

If Borges was the first author to import the quantitatively infinite into his writing, the twentieth-century graphic artist M. C. Escher

(1898–1972) may be fairly labeled the first artist to systematically explore the quantitatively infinite visually. Escher's *Cubic Space Division* provides a visual analogue of Borges' Library.

PLATE 3.1. *Cubic Space Division*, 1952. M. C. Escher (1898–1972).
© 2007 The M. C. Escher Company-Holland. All rights reserved.

Escher thought this print accurately represented what he took to be the infinitude of both space and time. He explained, "[The artist] must divide his Universe in distances of a specific length, in compartments that repeat themselves in endless series." And according to Escher, this subdivision represents the endlessness of time, "At every border crossing between one compartment and the next, [the artist's] clock ticks."[18] Although Aristotle did not believe space was infinite, Escher's print, and his explanation of it, illustrates an important feature of Aristotle's infinity that is addressed in the next section.

ARISTOTLE'S TWO INFINITIES

Aristotle rejected the existence an infinite material object, but he offered two ways in which an entity, whether it is material or not, could be infinite. In keeping with Aristotle's definition of an infinite entity, "not 'that which has nothing beyond itself' . . . but 'that which always has something beyond itself,'" these two conceptions of infinitude provide two ways in which, no matter how much of the entity is encompassed, or accounted for, some part of it would have been missed. This accounting of the entity associates the concept of infinite with process, perhaps with each tick of the clock, to steal an image from Escher. According to Aristotle, an entity is infinite if no matter how much of it has been listed, counted, accounted for, or contained (or imagined to have been listed, counted, accounted for, or contained) the entity will not have been entirely captured. The nature of the process used to obtain this accounting of the entity leads to Aristotle's two ways in which an entity can be infinite; according to Aristotle's *Physics* an entity can be *infinite by addition* or *infinite by division*.[19] These two forms of infinitude are most easily understood by considering mathematical objects.

A mathematical example of an entity that is infinite by addition is the collection of counting numbers, 1, 2, 3, and so forth. Aristotle knew what everyone else knew, that there is not a largest counting number (if you have a candidate add 1 to it). Since no list of counting numbers will itself be unending, any such list must necessarily be finite, containing only ten, ten thousand, or 10 billion entries. But no matter how many of the counting numbers have been listed, the list will not be complete; there will always be a counting number not on the list (for example the number obtained by adding all of the numbers on the list).

The other way Aristotle conceived something could be infinite is that it could be subdivided endlessly. A mathematical example of such an object is a line segment connecting two points A and B. (We examine challenges to this conception of a line segment in chapter 10.) It is possible to divide the line segment into two equal parts, and then to divide one of these two pieces into two equal parts, and to continue this process indefinitely. The intuition here is that any line segment, no matter how short, can be divided into two pieces, and so the process will never terminate.

Buried in the last sentence is again the idea of process, which Aristotle deftly used to avoid falling into the apparent trap set by the following example. Every other counting number is even—the even numbers are of course 2, 4, 6, 8, and so forth. Another way to view this list of even numbers is as $2 = 2 \times 1, 4 = 2 \times 2, 6 = 2 \times 3, 8 = 2 \times 4$, and so forth.

This way of representing the even numbers reveals the mathematical observation

Every counting number can be used to produce a
unique even number.

To obtain an even number just multiply the given counting number by 2. This means that the list of consecutive counting numbers and the consecutive even numbers are related by what medieval philosophers called a *correlation correspondence* and modern mathematicians call a *one-to-one correspondence*:

counting numbers: 1 2 3 4 5 6 7...
even counting numbers: 2 4 6 8 10 12 14...

If the first list, the list of all counting numbers, could be completed then it would be an unending, infinite list with two contradictory properties:

1. It would appear to be of the same size as the collection of even counting numbers (because of the one-to-one correspondence).
2. It would contain the list of all even counting numbers.

So, assuming the existence of the complete, infinite list of counting numbers yields two infinite lists, one of which is a *part* of the other. This appears to contradict the basic tenet that

the whole is greater than the part

since both the whole and the part are infinite, and so assumed to be the same.

Aristotle's solution was not to accept this part-being-as-large-as-the-whole situation. Instead, he did not allow the existence of the complete list of all counting numbers—the process of accounting for the counting numbers will never be completed. Aristotle did not say that such a collection was infinite, but that it was *potentially infinite by addition*.

The analogous situation holds for the infinite divisibility of a line segment. Suppose a line segment, connecting two points A and B, is divided into two pieces, at a point C, and then each of the two pieces is divided into two pieces. Further suppose that this process is continued until it stops. Then the unlimited number of divisions of the segment

AC are among the unlimited number of divisions of the entire segment *AB*; thus, one infinity is contained in another. As before this violates the principle the whole is greater than the part. Thus, Aristotle did not allow this subdivision to continue until it has been done an unlimited number of times; it was just a process that never stopped. As much as this conclusion confused many medieval thinkers, Aristotle concluded not that a line segment contains an infinite number of subdivisions but that a line segment could be subdivided indefinitely. Using Aristotle's terminology, a line segment is *potentially infinite by division*.

TIME AND SPACE—INFINITUDE IN THE MATERIAL WORLD

Of quantities some are discrete, others continuous. . . . Discrete are number and language; continuous are lines, surfaces, . . . time and place.
— *Aristotle,* Categories *(4th century B.C.)*

Since Aristotle did not equate the material and mathematical worlds, mathematical evidence did not necessarily tell him whether anything in the material world of human existence could be infinite. To investigate whether or not a material object, including a perhaps-evolving universe, could be infinite by addition, Aristotle applied one of his physical theories and one of his philosophical principles. Aristotle's understanding of existence depended on the idea that that which can be conceived of as existing must exist; existence is a property of being. So, if a material entity or collection could be infinite by addition then it would have to already exist as a completed whole. Thus, according to Aristotle's thinking, neither the universe nor the Library in Borges' short story could be indefinite; each would have to be either finite or infinite. If either the universe or the Library were to exist as a complete infinite structure, its existence would violate Aristotle's assertion that something could be infinite only if no matter how much of it has been delineated some part will have been left out. Thus Borges' Library could not exist, and a material entity could not be infinite by addition. In particular, neither the universe nor anything in it could be infinite by addition.

However, Aristotle maintained that both space and time were infinite by division, and he forever linked the infinite divisibility of the two by the following simple argument from his *Physics*.[20] Consider two moving objects A and T, A being the faster of the two. In the time, t_1, that A moves a given distance, d_1, T will move a shorter distance, d_2. And in the shorter time, t_2, that A moves the distance, d_2, T will move a still shorter distance, d_3. And so on, ad infinitum. It follows that if there were a shortest distance, then there would be a shortest increment of time, and if there were a shortest increment of time, then there would be a shortest distance. So just as there are shorter and shorter intervals of time there must be shorter and shorter distances in space. (Aristotle would have put this in precisely the opposite way, because for him time did not exist without motion. Since motion at a constant velocity can be imagined to be over as short a distance as desired, time can be imagined to have as short an interval as desired.)

Because Aristotle argued that a void could not exist, matter and space were coincident; since space is potentially infinite by division, so is matter. An immediate consequence of this is the rejection of any sort of an atomic theory; Aristotle's conception of the infinite forces him to reject both the Leucippus/Democritus and the Platonic theories of matter.

In several other prints, for example in *Smaller and Smaller*, below, Escher represented infinity by suggesting the infinite divisibility of space.

PLATE 3.2. *Smaller and Smaller*, 1956. M. C. Escher (1898–1972).
© 2007 The M. C. Escher Company-Holland. All rights reserved.

In *Smaller and Smaller* the outer ring of lizards fills half of the space between the center of the print and its outer edge; the next ring of lizards fills half of the space between the center of the print and the inside edge of the outer ring of lizards. This halving process is repeated with the next ring of lizards, and then the next. This stepwise process will never completely fill the canvas because there will always be a gap between the innermost ring and the center of the print.

Escher felt that this "centripetal reduction is . . . unsatisfactory because of the arbitrary outward limitation."[21] To overcome this, Escher experimented with reversing the pattern from *Smaller and Smaller* in a series of prints, having the larger images in the center of the circle and reducing them in size as they approach the boundary of the circle. See, for example, Figure 8.1 in the discussion of non-Euclidean geometry.

ZENO OF ELEA

[The] point of philosophy is to start with something so simple as not to seem worth stating, and to end with something so paradoxical that no one will believe it.
— *Bertrand Russell, "The Philosophy of Logical Atomism" (1918)*

Embedded in the link between the infinite divisibility of space and the infinite divisibility of time is Aristotle's response to one of the most subtle proponents of Parmenides' view that reality is a unified, homogeneous whole beyond our perception. This proponent was Zeno of Elea (born c. 490 B.C.), who appealed to presumed properties of mathematical infinity to prove Parmenides' assertion that reality must be a homogenous, metaphysical infinity. In providing his analysis of the nature of reality, two centuries before Aristotle rejected the metaphysically infinite, Zeno had further intertwined the metaphysically and mathematically infinite.

Each of Zeno's demonstrations purports to show that motion is impossible, and because we observe and experience what we perceive as motion, the world of our perception is an illusion. The real reality must be a unified, static whole beyond what we can ever know. What is remarkable about Zeno's so-called paradoxes is that whether space or time is assumed to be infinitely divisible or not, that is, continuous or discrete, one of them will seem to establish the impossibility of physical motion.

The first of Zeno's best-known paradoxes is called "the Dichotomy":

To get from Point A to Point B you first have to go half of the way, and then half of the remaining distance, and then half of the new remaining distance, and so on. You can never reach Point B because at any moment you are some fixed distance away from Point B and you have to travel half of that remaining distance before you can travel the entire remaining distance.

The Dichotomy supported Parmenides' distinction between the perceived world and *what-is* in two ways. First, this paradox was not intended to demonstrate that the motion we experience is impossible, but that motion is an illusion, because the physical act underlying what we perceive as motion is impossible. Motion appears to occur, but cannot, so reality and appearance are not the same.

The second, subtler, way in which the Dichotomy supports Parmenides' philosophy resides in what it tells us about the nature of *what-is*. The Dichotomy shows that Parmenides' metaphysical infinity must be static, not only because motion requires a transition from an existing state to a nonexisting state, but because motion requires the completion of an endless number of tasks. Thus, Zeno establishes a property of metaphysical infinity by appealing to our naive notion of mathematical infinity—that it is unattainable.

If the moral of the Dichotomy can be summarized as

> motion is impossible because it requires the movement to each of an unlimited number of places

then it appears to depend on the assumption that space is continuous, and so there can be no end to the possible positions between any two positions.

Aristotle examined the Dichotomy under his assumption that both time and space are infinite by division. Aristotle pointed out that the conclusion of the Dichotomy rests on three premises:

1. To move from any place to any other place, it is necessary to move half of the way first.

2. The number of these half steps is endless.
3. It is impossible to complete an endless number of tasks.[22]

Aristotle accepted each of these premises, but resolved Zeno's paradox by examining the very nature of motion. Since, for Aristotle, time was infinite by division, there cannot be two adjacent instants; we cannot compare the position of a moving object from one moment to the next. Thus, motion does not occur at an instant, but over an interval of time. Similarly, an object cannot be at rest at a single moment, but only over an interval of time. So, in the Dichotomy, movement from A to B does not require movement to each of the "halfway" points, as the first premise might lead you to believe. Indeed, for Aristotle, when something is in motion it is never at a position; for us to be able to say that something is in a particular position, it must remain there over an interval of time and so be at rest and not in motion. This means that movement from A to B does not involve motion to each "halfway" point, but movement *through* each "halfway" point.

Another paradox Zeno offered has become known as "Achilles and the Tortoise":

Suppose Achilles is going to race a tortoise, and to make it a fair race the slower tortoise is given a head start. When Achilles starts to run, the tortoise will be at some position along the track. When Achilles reaches the place where the tortoise was at the beginning of the race, the tortoise will have moved a little bit, and by the time Achilles reaches that point the tortoise will have moved a bit farther. Achilles can never reach the tortoise because every time he arrives at the place where the tortoise was the tortoise will have moved ahead a bit farther.

The original paradox, as reported by Aristotle, is less prosaic: "The slowest runner will never be caught by the fastest runner, because the one behind has first to reach the point from which the one in front started, and so the slower one is bound always to be in front."[23]

In his essay "Avatars of the Tortoise," Borges uses Achilles and the Tortoise to introduce the idea of *infinite regression*, more formally known as *regressus in infinitum*. Borges examines several philosophical applications of this principle, which were summarized by the novelist John Barth (b. 1930) in "The Literature of Exhaustion." According to Barth, Borges "carries through" the history of philosophical uses of infinite regression: "pointing out that Aristotle uses it to refute Plato's theory of forms, Hume to refute the possibility of cause and effect, Lewis Carroll to refute syllogistic deduction, William James to refute the notion of temporal passage, and Bradley to refute the general possibility of logical relations."[24] Borges could have also included a discussion of Kant's description of how we understand the magnitude of something. (Not all these applications are relevant to the main themes of this book.)

The infinite regression is not immediately evident in Achilles and the Tortoise, as this paradox is simply a restatement of the Dichotomy, with Achilles racing toward a moving target. Viewed this way Aristotle's resolution of the Dichotomy also resolves Achilles and the Tortoise. Another of Zeno's paradoxes, the Arrow, seems more apt. In Aristotle's version, the connection between infinite regression and the Arrow is still hidden: "If it is always true that a thing is at rest when it is opposite to something equal to itself, and if a moving object is always in the now, then a moving arrow is motionless."[25] The phrase "opposite to something equal to itself" should be interpreted as *it is in a space equal to its volume*.

Aristotle's examination of the nature of motion also resolved this paradox, but not its modern incarnation, below, which is the one Borges examined in his essay. Aristotle wrote that the conclusion that the moving arrow is motionless "depends on assuming that time is composed of nows."[26] If time consists of a sequence of adjacent moments, nows, instead of being infinitely divisible, then a moving arrow would have to be at one position at one moment, then in another position at the next moment. But as space is infinitely divisible, there are intermediate positions between the arrow's first and second positions, but the arrow can never occupy any of those positions because there is no intermediate time between two moments.

The modern reformulation of the Arrow, which is a reversal of the Dichotomy, more clearly depends on infinite regression:

> In order for an arrow to move from your bow to a target, it must first move to the halfway point between the bow and the target. Before

it can accomplish that motion, the arrow must first move halfway from the bow to the halfway point to the target. Yet again, before it can move to this second intermediate point it must move halfway from the bow to that point. Thus, the arrow can never be in motion, because before it can move to any point beyond its stationary position on the bow it must move one-half of that distance, and before that one-half of that distance, and so forth.

Aristotle did not address the Arrow in this form, whose conclusion also depends on assuming that time is made up of "nows." But even if time is infinitely divisible, it is not evident what happens to put a stationary arrow into motion.

Henri Bergson, one of the more modern proponents of the position that reality is metaphysically infinite, addressed this paradox. Bergson allowed for another way of knowing or exploring reality. According to Bergson, there are "two profoundly different ways of knowing a thing. The first implies that we move round the object; the second that we enter into it. The first depends on the point of view at which we are placed and on the symbols by which we express ourselves. The second neither depends on a point of view nor relies on any symbol."[27]

Bergson called knowledge obtained by intellect "relative," and knowledge obtained by intuition "absolute." The reality we can learn of only through intuition is Bergson's version of metaphysical infinity. Bergson turned to the Arrow to illustrate how an application of logical analysis, or intellect, could lead us astray. According to Bergson, continuity is a basic concept of reality, a concept that we understand completely through intuition. The mathematical, logical preoccupation with the concept of a *point* leads us to believe that continuity should somehow be built up from points. This is an example of the intellect imposing a condition on reality that it does not possess—the motion of the arrow in the paradox cannot be reduced to points. The arrow does not stop at intermediate positions, at intermediate times, according to Bergson (and Aristotle); instead the arrow's motion is a continuous, seamless flow. The analogy Bergson used was that of what was then called a cinematograph but to us is known as a motion picture projector.

According to Bergson, a mathematician analyzing motion, as movement from point to point, is attempting to understand the world by using the analogy of movie film. While watching a movie, for example of Achilles chasing a rabbit around a stadium track, we are watching a finite number of still photographs projected in sequence onto the screen at a rate of thirty frames per second. To us, Achilles' motion looks smooth—continuous. But of course it is not, it is a finite sequence of instances. Our eyes do not work instantaneously; it takes time (duration) for us to perceive an image. If a sequence of images is flashed onto the movie screen at a rate of one per second we feel as if we are watching a slide show. If the rate is increased to two images per second then to three images then to four, the slides' images remain on the screen for shorter and shorter intervals of time. There is some point at which we no longer perceive the individual slides as a static image. At this point, Achilles' motion appears to be continuous. Depending on several variables, for example, the brightness of the image, our visual acuity, Achilles' motion will evolve from jumpy to smooth at about twenty frames per second. Bergson said that imagining that motion occurs as observed in the cinematograph, even when the motion occurs in an unimaginably large number of frames per second, is imposing ideas from the unreal world of mathematics onto reality.

Neither Bergson's appeal to intuition nor an application of Aristotle's analysis of Zeno's other paradoxes resolves the Arrow. Both Aristotle and Bergson appealed, in slightly different guises, to the concept of continuity, but the Arrow depends on the discontinuity between being at rest and being in motion, and the seeming impossibility of making that transition. Confronted with having to explain how something at rest can be put in motion, and overcome the objection presented by the Arrow, a possibly exasperated Bertrand Russell (1872–1970) said, "When a body moves, all that can be said is that it is in one place at one time and in another at another. . . . Motion consists merely in the fact that bodies are sometimes in one place and sometimes in another, and that they are at intermediate places at intermediate times."[28] As evidence of the inconclusiveness of these philosophically inclined discussions of motion, we offer two twentieth-century pieces of art that seem to represent these contrary views. The first of these is Marcel Duchamp's painting *Nude Descending a Staircase (No. 2)*.

PLATE 3.3. *Nude Descending a Staircase (No. 2)*, 1912. Marcel Duchamp (1887–1968). Oil on canvas, 57⅞ × 35⅛ in (147 × 98.2 cm). The Louise and Walter Arensberg Collection, 1950. Philadelphia Museum of Art, Philadelphia. Photo: The Philadelphia Museum of Art / Art Resource, New York. © 2007 Artist Rights Society (ARS), New York / ADAGP, Paris / Succession Marcel Duchamp.

In this painting Duchamp illustrated motion by allowing time to be an element in his painting—it appears as if Duchamp were viewing motion as a sequence of discrete steps. Partly in response to Duchamp's painting, Umberto Boccioni (1882–1916) wrote in one of his manifestos on futurist art (1913): "It seems clear to me that this *succession* is not to be found in the repetition of legs, arms, and faces, as many people have idiotically believed, but is achieved through the intuitive search for the *one single form which produces continuity in space*."[29] Boccioni's sculpture *Unique Forms of Continuity in Space* seems to illustrate his view of motion as a continuous process.

PLATE 3.4. *Unique Forms of Continuity in Space*, 1913. Umberto Boccioni (1882–1916). Galleria d'Arte Moderna, Milan, Italy. Photo: Scala / Art Resource, New York.

As important as Zeno's paradoxes, and the philosophical examinations of them, were to the evolution of the mathematical ideas, theological discussions were even more significant. And as we see in the next chapters, these discussions examined not only the concept of infinity but also the properties of geometric objects and the meaning of proof.

4

The Flat Earth and the Spherical Sky

*[Is] there any one so senseless as to believe that there
are men whose footsteps are higher than their heads?
or that the things which with us are in a recumbent
position, with them hang in an inverted direction? that
the crops and trees grow downwards? that the rains,
and snow, and hail fall upwards to earth?*
— Lactantius, The Divine Institutes (early
4th century)

One morning, in the third century B.C., the Greek astronomer Eratosthenes (276–194 B.C.) did a seemingly curious thing; he stuck a straight stick into the ground, as perpendicular to the earth as he could, and watched its shadow shift as the sun moved across the morning sky. Eratosthenes was in Alexandria, Egypt, so the sun stayed in the southern sky as it arched toward its apex. At midday, Eratosthenes very carefully measured the angle between the tip of the stick and the tip of its shadow. With this single measurement, Eratosthenes knew he could determine the size of the earth.

Eratosthenes' calculation of the earth's circumference relied on an assumption about the mathematical properties of parallel lines. The parallel lines in this experiment are rays of sunlight, and although rays of sunlight are not exactly parallel, if the sun is very far from the earth then they are close to being parallel. To see this, imagine standing on the sun at the point from which two rays emanate. If you were to aim a very strong telescope at one spot on the earth, and then turn it to another spot on the earth, you would barely need to make any adjustment. The change in the direction of your telescope would be immeasurably small, and the smaller the angle between two lines the closer they are to being parallel. So, although it is not absolutely correct, Eratosthenes assumed for the sake of his calculation that two rays of sunlight traveling through space moved along parallel paths.

For Eratosthenes to have made this assumption, he needed to know that the distance from the sun to the earth is fairly large. If the sun were a close-by, small disc, Eratosthenes' assumption that rays of sunlight are essentially parallel when they strike the earth would be grossly incorrect. By just looking at the sky it is impossible to tell whether the sun is a small disc relatively close to earth or a large disc at a great distance. But almost no one thought the sun was very close to earth. As far back as the sixth century B.C., Anaxagoras had suggested that the sun was a disc twenty-eight times as large as his proposed cylindrical earth, and since it appears to be rather small in the sky, it must be relatively far away. However, Eratosthenes' assumption that the sun is not too nearby probably came from the work of his older contemporary Aristarchus (310–c. 230 B.C.), whose ingenious use of the properties of similar triangles led to estimates for the distances between the earth, moon, and sun. We briefly describe Aristarchus' work, if only to emphasize its geometric character, later in this chapter; first we explore the meaning of Eratosthenes' experiment.

THE FLAT EARTH

Another crucial component in Eratosthenes' experiment was his knowledge of a well in Syene, the modern Aswan, which is directly beneath the midday sun twice a year. Such a seemingly difficult-to-verify fact could be established because at the moment the bottom of the well is fully illuminated the sun is directly overhead. Eratosthenes also needed to know precisely when the sun reached its apex over Syene; since Syene is, more or less, due south of Alexandria, when the sun is directly above the well it is at its highest point in the Alexandrian sky. Under the assumption that the sun is very far from earth, the very existence of the shadow of the stick demonstrates that the surface of the earth is not flat.

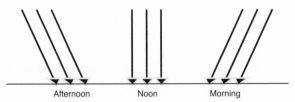

Afternoon   Noon   Morning

FIGURE 4.1. If the earth were flat and the sun far away, then at any time of day parallel rays of sunlight would strike the earth at the same angle.

So at midday, when Eratosthenes measured the angle, since the sun was directly over the well in Syene it would also be directly over his stick. Thus, the stick would not cast a shadow, and so there would not be an angle to measure; the existence of the shadow confirmed what was already believed—that the earth is not flat.

The Pythagoreans, Plato, and Aristotle had all believed the earth was a sphere. Although the belief that the earth is a sphere might be driven by the belief that the earth is a geometrically perfect, three-dimensional shape, there is visual evidence to support this belief. One piece of evidence comes from watching a ship as it moves away from port, toward the horizon. The ship appears not only to shrink in size, but to slowly sink into the sea. Another piece of evidence is more cosmological. If you believe, as the Greeks generally did, that the moon and the sun revolve around the earth, then it is easy to conclude that when there is a lunar eclipse the earth is between the sun and moon. And since whenever the moon enters the earth's shadow the shadow is circular, it is reasonable to guess that the earth is most likely a sphere.

In *On the Heavens*, Aristotle cited the shape of the earth's shadow on the moon during an eclipse to support his view that the earth is spherical, "in eclipses the outline is always curved: and, since it is the interposition of the earth that makes the eclipse, the form of this line will be caused by the form of the earth's surface, which is therefore spherical."[1] Aristotle continued with further physical evidence: A change in latitudes affects which stars can be seen in the night sky. He noted that different stars are visible when the night sky is viewed from Egypt and from "northerly regions," which means that the earth is not flat. Further, since different stars are visible if you make even a relatively small shift in your north-south position, the earth cannot be very large (at least when compared with the distances to the stars). Aristotle offered additional evidence that the earth is a sphere—the existence of elephants both to the west and east of Greece, "about the pillars of Hercules and the parts about India, suggesting that the common characteristic of these extremes is explained by their continuity."[2]

Despite the visual evidence and Eratosthenes' experiment, the belief that the earth is flat did not completely disappear; it did, however, become the minority view. Among the Greeks who play an important role in the history of mathematical ideals, only Democritus questioned the dominant view that the earth is a sphere. According to Aristotle, Democritus gave "the flatness of the earth as the cause of its staying still. ... This seems to be the way of flat-shaped bodies: for even the wind can scarcely move them because of their power of resistance."[3]

In the first millennium only a few writers argued that the earth was flat. Of particular relevance to the nature of proof are two Christian apologists, Lactantius Firmianus, early in the fourth century, and Cosmas Indicopleustes, in the sixth century. Between 303 and 311, Lactantius wrote what is considered to be his most important treatise, *The Divine Institutes*. In this work, Lactantius discussed what he perceived to be the flaws in pagan beliefs—that is, the errors of non-Christian writers, especially the Greeks. This work contains seven books; the titles of the first four books reveal part of its focus:

Book 1. Of the False Worship of the Gods
Book 2. Of the Origin of Error
Book 3. Of the False Wisdom of Philosophers
Book 4. Of True Wisdom and Religion

One of the ideas from pagan thought Lactantius sought to undermine, in book 3, was the belief that the earth is round, which allowed for the possibility of the other side of the earth being inhabited by antipodes. Lactantius asked whether anyone could be so senseless as to believe in the existence of these humans.

The sixth-century writer Cosmas Indicopleustes should have known that the earth is not flat. Early in his life Cosmas traveled extensively, there is evidence that he sailed both the Mediterranean and Red Seas; his surname ("India-voyager") suggests that he had been to India, so he should have been aware of the physical evidence cited by Aristotle. But later in life, Cosmas converted, or returned, to Christianity, began a monastic life, and defended his interpretation of the Bible against ancient, pagan beliefs in his *Christian Topography*.[4]

Cosmas' *Topography* was divided into twelve books, including:

Book 1. The Places and Figures of the Universe; Heresy of Affirming That the Heavens Are Spherical, and That There Are Antipodes; Pagan Errors as to the Cause of Rain and of Earthquakes
Book 4. A Recapitulation of the Views Advanced; Theory of Eclipses; Doctrines of the Sphere Denounced

Among Cosmas' arguments for the flatness of the earth was his attempt to counter the conclusion that the shape of the shadow on the moon during a lunar eclipse tells us something about the shape of the earth. His explanation for the roundness of the shadow of the earth depended on his description of the earth. According to Cosmas the earth is a flat rectangle with a high central mountain and the circular edge of the shadow that appears on the moon during a lunar eclipse is not caused by the earth itself but by the earth's central mountain.

Cosmas concluded with an appeal to belief as the ultimate source of truth: "To enquire further into these matters we have no leisure; for such knowledge is unprofitable to us who have access to a more profitable knowledge, which imparts to our soul a good and beneficent hope which God hath promised he will give to those who believe in him, while those who act unjustly he has doomed to perdition."[5] Later, Cosmas added book 6, "Regarding the Size of the Sun," wherein he defended his conclusions: "After my work had been finished, some questioned us about the figure of the world, saying: 'How can the sun possibly be hidden, as you hold, by the northern parts of the earth, which according to you are very high, while he is many times larger than the earth? For in the case of the sphere which we advocate, however much greater the sun may be than the earth, he will always, when giving light to one part of her surface, leave the other in shadow.' To those so questioning us we have made a very brief reply, that such a thing is false and a pure fiction."[6]

Cosmas then used the existence of the shadow in Eratosthenes' experiment to prove that the sun must be a small, nearby disc. His argument is that because the earth is flat, if the sun were far away, then at any moment, all shadows would be cast at the same angle. Since this is not the case, the sun must be so close to the earth that rays of sunlight are not nearly parallel. Eratosthenes' argument can be paraphrased as: Since the sun is far away, the existence of the stick's shadow in Alexandria, when the sun is directly over the well in Syene, demonstrates that the earth is not flat. Cosmas' argument can be paraphrased as: Since the earth is flat, the existence of the stick's shadow in Alexandria, when the sun is directly over the well in Syene, demonstrates that the sun must be close to the earth. As Figure 4.2 illustrates, both Cosmas' and Eratosthenes' arguments are correct; their conclusions are consistent with their different assumptions.

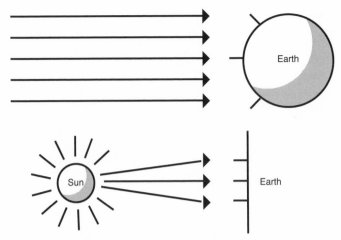

FIGURE 4.2. Both Eratosthenes' and Cosmas' assumptions about the shape of the earth and the distance from the earth to the sun are consistent with shadows being cast at different angles by three posts that are placed far apart.

Cosmas also used the differences in the lengths of these shadows to argue that the sun is considerably smaller than the earth. Once he had established that the sun is small and nearby, the conclusion that the shadow of the earth's central mountain could cause an eclipse is reasonable.

In his book *Inventing the Flat Earth* (1991) Jeffrey Burton Russell examined the origins of the belief that during the Middle Ages the earth was thought to be flat. Neither Lactantius nor Cosmas was widely read in Europe, so Russell does not trace this modern assumption back to them. None of Cosmas' *Topography* was translated into Latin until 1706, and Lactantius was all but ignored until the Renaissance. However, there is one important reference to Lactantius, in Copernicus' *On the Revolutions* (1543). Copernicus knew the church would not welcome his proposition that the earth revolves around the sun, so he wrote a preface to Pope Paul III. In the preface Copernicus surveyed the history of the heliocentric universe and warned, "there will be babblers who claim to be judges of astronomy although completely ignorant of the subject." A few lines later he gave Lactantius as an example: "It is not unknown that Lactantius, otherwise an illustrious writer but hardly an astronomer, speaks quite childishly about the earth's shape."[7]

According to Russell, many of the early modern references to the belief that the earth is flat appear in fiction. For example, in Somerset Maugham's *Of Human Bondage* (1915) one of the characters says, "Saint Augustine believed that the earth was flat and the sun turned around it." However, Augustine seemed to accept that the earth might be a sphere; he just was not convinced that this meant that the entire earth was inhabited. As Augustine put it, "As for the fabled 'antipodes,' ... even if the world is supposed to be a spherical mass, or if some rational proof should be offered for the supposition, ... even if the land were uncovered, it does not immediately follow that it has human beings on it."[8] After reiterating that human existence began at a single place on earth, Eden, Augustine continued that "it would be too ridiculous to suggest that some men might have sailed from our side of the earth to the other ... so that the human race should be established there."[9]

Another fictional reference to the belief that the earth is flat appeared in Washington Irving's *Life and Voyages of Christopher Columbus* (1828). Irving wrote that before Columbus left on his voyage, he met with a group of theologians, and inquisitors, in Salamanca, Spain. According to Irving, this group argued that the world is flat. Probably no such meeting was ever held and no such group ever existed.

EUCLID'S ELEMENTS

Eratosthenes' experiment depended on a special property of parallel lines. Both to understand Eratosthenes' experiment, and to appreciate the impact of the principles of Greek geometry on theology, art, and later mathematics, it is important to have at least a passing familiarity with Euclid's *Elements* (3rd century B.C.).

Geometry, as it is presented in the *Elements*, is a deductive system. Its assumptions about geometric objects, and the relationships between them, are stated explicitly; additional properties of geometric objects are deduced from those assumptions using the commonly accepted rules of logic. Among these logical rules are the generally accepted syllogisms, for example the well-known *modus ponens*

Socrates is a man.
All men are mortal.
Therefore Socrates is mortal.

Euclid began book 1 of his *Elements* with twenty-three *definitions*, five *common notions*, and five *postulates*. Euclid's list of definitions begins:

1. A point is that which has no part.
2. A line is breadthless length.

and ends with

23. Parallel straight lines are straight lines which, being in the same plane and being produced indefinitely in both directions, do not meet one another in either direction.

Euclid's five common notions are the basic principles of mathematics; they include at least one assumption that has already played a significant role in our review of mathematical ideas in theology:

1. Things that equal the same thing also equal one another.
2. If equals are added to equals, then the wholes are equal.
3. If equals are subtracted from equals, then the remainders are equal.
4. Things that coincide with one another are equal to one another.
5. The whole is greater than the part.

The five postulates of Euclidean geometry are statements about the properties of geometric objects whose truths are supposed to be based on observation or intuition. Euclid's first three postulates mirror our expectations for how a line or circle could be constructed:

1. It is possible to draw a straight line from any point to any point.
2. It is possible to extend a finite straight line continuously in a straight line.
3. It is possible to describe a circle with any center and radius.

Euclid's fourth postulate,

4. All right angles are equal to one another

may seem like a theorem rather than an assumption, but Euclid did not want to use angular measurement, at least in part because the Pythagorean discovery of incommensurable lengths had made the concept of measurement suspicious. So for Euclid, a right angle is not defined as being a 90-degree angle but an angle formed when two lines cross in such a way as to produce four equal angles. So, according to Euclid's fourth postulate, all angles formed in this way are equal.

Euclid's final postulate is less simply stated, and over time it came to be seen as being either unnecessary or false. This is Euclid's *parallel postulate*, and it gives a criterion for when two lines are parallel:

5. If two lines, L_1 and L_2 are crossed by a third line [as in Figure 4.3] and if angle no. 1 and angle no. 2 add up to less than two right angles [less than 180 degrees in our language] then the lines L_1 and L_2 will cross when they are extended sufficiently far to the right.

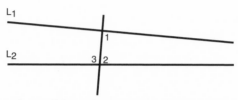

FIGURE 4.3. The parallel postulate states that if angle no. 1 + angle no. 2 does not equal 180 degrees, then the lines L_1 and L_2 are not parallel.

The intuition behind this postulate is clear, if we imagine that in Figure 4.3 angle no. 2 equals 90 degrees and angle no. 1 is less than 90 degrees, this postulate then asserts that the two lines L_1 and L_2, if extended sufficiently far, will cross somewhere off to the right.

This postulate owes its awkwardness to the Greek avoidance of infinity—a line is infinite by addition in that it can be extended indefinitely but it is not an existing infinitude. Using only the definition of parallel lines, it is not possible to determine whether or not two particular lines are parallel—just because two lines have been extended and found not to yet intersect does not mean they will not intersect if they are extended further. It would take the completion of an infinite process, the complete extension of two lines, to determine that two lines never meet, and so are parallel. The ingenuity of Euclid's parallel postulate, and its companion theorem that if angle no. 1 and angle no. 2 add to 180 degrees then the lines L_1 and L_2 are parallel, is that it reduces the question of whether or not two lines are parallel to the examination of two angles.

The geometric property of parallel lines Eratosthenes called upon in his experiment is a consequence of Euclid's parallel postulate. This is because if the lines L_1 and L_2 are parallel then, in our language

angle no. 1 + angle no. 2 = 180 degrees, and
angle no. 2 + angle no. 3 = 180 degrees

from which it follows that angle no. 1 = angle no. 3. We will describe how Eratosthenes exploited this equality, once we understand how he knew the sun was very far from earth, and so rays of sunlight are, essentially, parallel.

WHY THE SUN IS FAR AWAY

There are some, King Gelon, who think that the number of the sand is infinite in multitude; and I mean by the sand not only that which exists about Syracuse and the rest of Sicily but also that which is found in every region whether inhabited or uninhabited.... I will try to show you by means of geometrical proofs, which you will be able to follow, that, of the numbers named by me ... some exceed not only the number of the mass of sand equal in magnitude to the earth filled up in the way described, but also that of a mass equal in magnitude to the universe.... Aristarchus of Samos brought out a book consisting of some hypotheses, in which the premises lead to the result that the universe is many times greater than that now so called.

— Archimedes, "The Sand Reckoner" (3rd century B.C.)

In the above excerpt from "The Sand Reckoner," in which Archimedes (c. 287–c. 212 B.C.) devised a scheme for representing large numbers using the limited Greek symbols, Aristarchus is said to have proposed that the universe is quite large. The premises Aristarchus used to reach this conclusion were that the earth revolves around the fixed sun and that the center of the sphere of stars is at the sun. Thus, as the earth moves around the sun it is sometimes closer to the stars than at other times. However, the stars never seem to change in size. From this it follows that the stars are very far from earth.

Aristarchus thought not only that the stars are very far away but also that the earth is not very close to the sun. To reach this conclusion, Aristarchus employed simple geometric principles and an ingenious geometric experiment. It was well known that the moon is closer to the earth than the sun is, because, for example, the moon occasionally passes between the earth and sun. And Aristarchus understood that the relative positions of the sun, moon, and earth caused the various phases of the moon, and that when exactly one-half of the moon is illuminated, at either the first quarter or third quarter, the positions of these three bodies can be diagramed as:

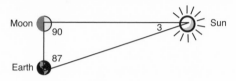

FIGURE 4.4. The triangle behind Aristarchus' calculation of the relative distances of the earth to the moon and the earth to the sun.

At the moment precisely one-half of the moon is illuminated, when viewed from earth, the angle at the moon will equal 90 degrees. If, at that moment, an observer on the earth were to measure the angle at the earth, formed by looking at the sun and then at the moon, it would be possible to use similar triangles to find the relative distances from the earth to the moon and from the earth to the sun. Aristarchus measured this angle and found it to be, roughly, 87 degrees (as indicated in Figure 4.4). From this one measurement, and properties of similar triangles, Aristarchus was able to conclude that the sun is between eighteen and twenty times farther from the earth than the moon is.

The earth-to-sun distance is now known to be roughly 390 times as great as the earth-to-moon distance, and Aristarchus could have discovered a similar, larger estimate because there are not any theoretical flaws in his reasoning. If it had been possible for Aristarchus to exactly measure the 87-degree angle, at precisely the moment one-half of the moon is illuminated, he would have found that it is approximately 89½ degrees. Alas, that precise moment can never be determined because the moon is not a perfect sphere, and the measurement of the angle will never be exact, but will necessarily fall into a small range of possibilities.

Aristarchus' estimate for the ratio of the earth-to-moon and earth-to-sun distances shares a feature with almost all ancient, medieval, and Renaissance cosmic measurements—it is relative. Just knowing that the sun is eighteen to twenty times farther away from earth than the moon does not give the distance to either body. If the moon is fifty miles from earth, then the sun is between 900 and 1,000 miles from earth. On the other hand, if the moon is 10,000 miles from earth then the sun is between 180,000 and 200,000 miles from earth. In the first case the sun is fairly close to earth, and Eratosthenes' assumption that rays of light travel from the sun to the earth along parallel paths is measurably incorrect. Fortunately for Eratosthenes, Aristarchus did not stop with his estimation of the relative distances of the moon and sun from the earth. By studying the shadow of the earth during a lunar eclipse and the shadow of the moon during a solar eclipse, Aristarchus was able to attach values to these distances and to the size of the sun and moon. For example, he estimated that the distance to the sun was 780 earth-diameters and the diameter of the sun was 7 earth-diameters. Just from knowing that ships had already traveled great distances, Eratosthenes knew the earth must be large in human terms; so, using Aristarchus' estimate, that the sun is 780 earth-diameters away, he knew the sun is very far from earth.

ERATOSTHENES' EXPERIMENT

Eratosthenes' calculation was based on the Greek understanding of the behavior of parallel lines—that when two parallel lines cross another line, as below, the indicated angles, α and β, are equal.

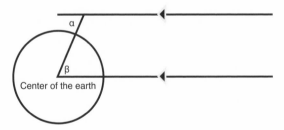

FIGURE 4.5. A schematic representation of Eratosthenes' experiment.

The stick Eratosthenes stuck in the earth was perpendicular to the earth's surface; so, if it were to be extended into the interior of the earth it would pass through the earth's center. Similarly, if the ray of sunlight striking the well at Syene were extended into the earth, it too would pass through the earth's center. In Figure 4.5 the top horizontal line represents the ray of sunlight that hits the stick at Alexandria and produces the stick's shadow; the bottom horizontal line represents the ray of sunlight that is perpendicular to the surface of the earth at Syene; and the slanted line represents Eratosthenes' stick (the last two lines are drawn as if they extend to the center of the earth).

Eratosthenes needed to calculate the angle β, which represents how much of the circumference of the earth is contained in the portion from Syene to Alexandria. But Eratosthenes could not measure the angle β directly, as it is located at the center of the earth. Instead, on the appropriate day, at the appropriate moment, Eratosthenes measured the angle α, which, because of the assumed properties of parallel lines, equals the angle β.

Using his estimate for the angle α, which in our language was 7½ degrees, Eratosthenes concluded that the ratio of the distance from the well to his stick to the circumference of the earth equaled the ratio of 7½ degrees to 360 degrees. This ratio yields an estimate for the circumference of the earth in the Greek measurement system of 50 × 5,000 stades = 250,000 stades, just over half as large as the 400,000 stades Aristotle claimed for the earth in *On the Heavens*. We are not certain how to convert a Greek stade into feet, but a reasonable estimate is that one stade equals just a bit more than 600 feet.[10] Using this estimate, Eratosthenes would have calculated the circumference of the earth as being 28,732 miles. This is at least on the same scale as the current estimate of just under 25,000 miles.

A SHORT OVERVIEW OF THE GEOMETRIC COSMOS

From man or angel the great Architect
Did wisely conceal, and not divulge
His secrets to be scann'd by them who ought
Rather admire; or, if they list to try
Conjecture, he his fabric of the heavens
Hath left to their disputes; perhaps to move
His laughter at their quaint opinions wide
Hereafter, when they come to model heaven
And calculate the stars; how they will wield
The mighty frame; how build, unbuild, contrive,

To save appearances; how grid the sphere
With centric and eccentric scribbled o'er,
Cycle and epicycle, orb in orb.
— *Milton,* Paradise Lost *(1667)*

It is no more reasonable to treat the heavens as a single entity than it is not to differentiate stones from turtles. And the longer you look at the heavens, the subtler they become. Anyone observing the sky for a few days will notice that it contains three types of objects: the sun, the moon, and the multitude of stars. Because the moon obscures the sun during a solar eclipse, the moon must be closer to the earth than the sun, and because at sunrise and sunset the backdrop of stars is behind the sun, the sun must be closer to the earth than the stars. This at least provides an ordering—moving away from earth you would encounter the moon then the sun and then the stars.

At first, all of the stars appear to move in unison, maintaining their relative positions as they cycle across the sky. Beyond the sun there appears to be a single realm. But a more patient observer, someone watching the sky for a few months, will notice that occasionally a star will break rank. When this happens, the rogue star seems to have an independent motion; it is as if it has free will. By watching the night sky over a longer period of time, for years instead of months, one recognizes that there are only five different wanderers (planets) and that they follow different, mathematically imperfect paths. Distinguishing these five planets from the backdrop of stars yields a universe with its nine observable bodies: earth, sun, moon, five planets (Mercury, Venus, Mars, Jupiter, and Saturn), and the fixed stars.

In Aristotelian physics, a natural motion is any motion free of interference, for example the motion of a freely falling body, the sun or, presumably, a comet. The reason natural motions occur, according to Aristotle, is that every body in the universe has a natural place; that is, every body has a preexisting and permanent preferred position. Aristotle even offered a not-at-all-enlightening operative definition of a body's place in his *Physics*:

1. The tendency is for the place of a body to contain the body.
2. The place of a body is the position from which any motion of the body is to be measured.[11]

Aristotle appealed to the obvious, at least to him, truth that the universe seeks order and so any body not in its place will naturally move toward it.

As mathematical perfection was an unacknowledged requirement for the acceptance of an idea as a truth, any natural motion was either in a straight line or around a circle. Aristotle's two principles defining the place of a body imply that if a body is not in its place it will move toward its place either along a line or along a circle, and the movement of a body can only occur when it is not in its place. Each of these types of motion requires a center—a natural linear motion is either toward or away from a center and a natural circular motion is around the center of the circle.

Of all of these natural motions, one was assumed to be superior and, thus, necessarily the one in the heavens. In a short passage in the *Timaeus* Plato had written that the Creator allotted to the universe the one of the seven physical motions "which most properly belongs to intelligence and reason," which, according to Plato, was uniform circular motion.[12] The circle was thereby ensconced as the most geometrically perfect form—the one associated not with earthly matter but with the fabric of the cosmos—and circular motion displaced the six types of straight motion—up and down, forward and back, right and left—as the most divine.

The only way Aristotle could reconcile the movement of the sun around the earth and the movement of a released stone toward the ground, was to propose that there is a single center for all natural motions, necessarily at the center of the earth. Thus, all celestial motions had to be in perfect circles around the center of the earth. This reasoning led Aristotle to abandon the mathematical perfection of the Pythagorean cosmos, consisting of ten bodies moving in perfect circles around the central fire. Aristotle replaced the Pythagorean description of the universe with an earth-centered and highly problematic one.

It is another tenet of Aristotle's physics, and his theory of motion, that all natural motions are uniform. And, he argues, "Only circular movement can be continuous and eternal."[13] Thus all celestial motions are uniform, circular ones. The most glaring challenge to this claim is the irregularity of planetary motions. During their movement across the sky three of the planets, Mars, Jupiter, and Saturn, occasionally stop and reverse direction, move backward for a few days or months against the backdrop of stars, then again reverse direction and resume their original course. Eventually each of these planets traverses the sky and then disappears for a few or many months before reappearing. The other two planets, Mercury and Venus, are more obedient. They never reverse direction and they never move far from the horizon, but they never traverse the sky. These observed behaviors mean that either the geometry of the planetary paths, or the speed and direction of their motions, is not in accord with Aristotle's Pythagorean physical principles. Rather than abandon the mathematically perfect model for the universe, with all heavenly bodies moving along perfect earth-centered circles with uniform velocities, the Greeks sought a mathematical solution to the conflict between observed and theorized planetary behavior.

To understand how the Greeks reconciled observation with theory, and so truth with their assumptions about the role of beauty in the uncovering of truth, it is important to recall that for them there was no vacuum in space. Celestial objects were not imagined to be moving through emptiness, but through the material heavens, so it was not unreasonable to postulate that the sun, moon, planets, and stars were attached to spheres that revolved around the center of the earth. But which heavenly bodies belong to which spheres? The Greeks reasoned that the slower a planet appears to move across the sky, and so the longer it takes to traverse the backdrop of stars, the farther it is from earth. This is an entirely reasonable assumption if you believe that mathematical perfection requires the planets to all move at the same speed. Prolonged observation reveals that a complete orbit of Saturn, Jupiter, or Mars, requires thirty, twelve, or two years, respectively. This led the Greeks to believe that Saturn is the farthest planet from earth, beyond the orbit of Jupiter, which is beyond the orbit of Mars.

The length-of-orbit approach fails when applied to Mercury or Venus because they are always near the horizon, always within a few degrees of the sun. Aristotle adopted the sequence from Plato's *Republic*, putting Venus closer to earth than Mercury. This gives the sequence of bodies (listed from closest to earth to farthest from earth):

moon, sun, Venus, Mercury, Mars, Jupiter, and Saturn.[14]

Beyond Saturn is an outermost sphere of fixed stars. This celestial sphere appears to rotate on an axis passing through the earth's north and south poles, which accounts for the nightly procession of the backdrop of stars across the sky. The moon, sun, and planets are all contained within this bounded world, and each of them belongs to its own sphere rotating around the earth.

In this view, the universe consists of a sequence of concentric spheres, centered at the earth. Unfortunately this mathematically pure model, along with the assumption that all celestial motions are uniform (or at least regular), does not explain the wanderings of the planets. There were successive elaborations of this nested-spheres model. The first two were offered by Eudoxus, a student of the Pythagorean Archytas, and by Aristotle, and the third by the astronomer Ptolemy in the second century A.D. Eudoxus proposed an ingenious solution to the problem of the wandering planets: The motion of each planet is determined by the rotation of more than one sphere. The idea is that for each planet, there is a system of concentric spheres, with the planet attached to one of these spheres; these spheres rotate around the earth, in such a way that the resulting motion matches observation. By positing the existence of twenty-seven spheres, Eudoxus accounted for all observed planetary motions.

Eudoxus was at least partially Platonic in his outlook so he thought of this system of concentric spheres only as a mathematical construction that described the observed heavenly motions. The spheres were not material; they were as idealized as are all objects in the other, Platonic world. Because Aristotle had already rejected Plato's otherworldliness, when he adapted Eudoxus' model to fit with his physics, he conceived of these spheres as invisible material bodies.

Aristotle's universe contained thirty-three earth-centered concentric spheres: four for Saturn, four for Jupiter, and five each for the moon, Venus, Mercury, the sun, and Mars. This physical system needed a mechanism for all of its motion, and the mechanism Aristotle offered was the rotation of the celestial sphere of the stars by the unmoved mover. As the celestial sphere rotates, it forces the smaller spheres that control the motions of the sun, moon, and planets to rotate as well. To accomplish this transfer of motion, Aristotle postulates the existence of an additional twenty-two unrolling spheres that move the spheres associated with each heavenly body. The image to have is of rotating gears that cause other gears to rotate.

This model could not accommodate either the increasingly accurate records of planetary motions (especially for the motions of Venus and Mercury) or the slow movement of the celestial sphere. Thus, in the second century, Ptolemy (83–161 A.D.) introduced yet another structure of rotating mathematical spheres to explain the heavens. Ptolemy adapted an idea attributed to the Greek astronomer Hipparchus (second century B.C.) to give a more accurate accounting for planetary motions. In this scheme, each perfectly spherical heavenly body is attached to a sphere whose center is in the celestial region, called an epicycle, and as the epicycle rotates around its center, that center moves around earth in a perfect circle.

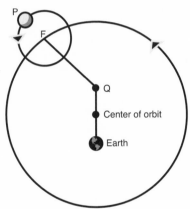

FIGURE 4.6. An illustration of the geometry behind Ptolemy's epicycles.

This is an amazingly sophisticated system, having one scheme for Mercury and another for the other four planets, the sun, and the moon. Yet for each heavenly body, all of these motions are uniform in that the rate of change in the measurement of angle at the point Q is constant. This accounts for the apparent changes in the direction and speed of the planets through an application of harmonious Pythagorean principles but does not conform to Aristotle's theory of natural motion. Equally as important, the movements of this system cannot be determined by a system of real, or imagined, sphere-gears, as was Aristotle's.

This is but one of the problems with Ptolemy's model for the universe that theologians and natural philosophers struggled with in the thirteenth through seventeenth century. Just as Cosmas' conclusion that the sun is not too far away can be deduced from his assumption that the earth is flat, theological solutions to conflicts between observation and belief are plausible when seen as part of their axiomatic system.

5

Theology, Logic, and Questions about Angels

It is therefore better that a proposition which cannot be demonstrated be received as an axiom, or that one of the two opposite solutions of the problem be accepted on authority.

— *Maimonides,* Guide for the Perplexed *(12th century)*

In the sixth century, the poet, politician, philosopher, and perhaps martyr Boethius (c. 480–c. 525), who had already translated portions of Euclid's *Elements* into Latin, introduced a variant of the axiomatic method to theology. In one of his writings, "How Substances Are Good in Virtue of Their Existence without Being Substantial Goods," which is known as "De hebdomadibus," Boethius provided an axiomatic foundation for his arguments. In the prologue, in order to "explain a little more clearly," Boethius wrote: "I have put forward first terms and rules on the basis of which I will work out all the things that follow, as is usually done in mathematics."[1] Boethius then presented nine axioms that were intended to support his theological conclusions. Two examples will suffice to illustrate the nature of these axioms.

Boethius' first axiom explains that a self-evident truth is "a statement that anyone approves once it has been heard." Of these self-evident truths Boethius delineated two types: those that rely on common sense (Boethius gave the example: "If you take away equals from two equals, what remain are equals") and those that "the learned but not the uneducated acknowledge" (Boethius' example: "Things that are incorporeal are not in a place"). Boethius' second axiom is an example of the latter type, "Being and that which is are different."[2]

The reason we said that Boethius introduced a *variant* of the axiomatic method to theology, rather than that he introduced the axiomatic method to theology, is because he did not attempt to deduce his conclusions from his axioms. Rather, Boethius explained that "a careful interpreter" could reach the same conclusions as he reached through an application of the detailed deductive reasoning he did not provide.

Boethius may be viewed as the first in a distinguished line of Christian thinkers who maintained that the deductive method, or logic, played a role in discussions of faith. This line included Anselm of Canterbury (1033–1109), Peter Abelard (1079–1142), and to a lesser extent, Peter Lombard (c. 1095–1160). Each of these theologians appealed to logic or deductive analysis differently. For example, Anselm gave his definition of God as that "than which nothing greater can be conceived" in his *Proslogion* (1074). (In Anselm's words, "even the fool is convinced that something exists in the understanding, at least, than which nothing greater can be conceived.")[3]

Abelard, who is known for his entanglement with Heloise that led to his castration, is considered to be one of the profoundest thinkers of his era. Modern philosophers still remark upon the originality and insightfulness of his writings on logic, ethics, and metaphysics, especially his theory of universals. Here we are interested in his approach to theology. Abelard did not believe that the conclusions of theologians should go unexamined. To illustrate why he believed this, he wrote his influential *Sic et Non* (1120), in which he listed a large number of propositions, such as "that to God all things are possible," and offered theological arguments both for and against the proposition. Abelard stated his purpose in writing *Sic et Non* in its prologue: "There are many seeming contradictions and even obscurities in the innumerable writings of the church fathers. Our respect for their authority should not stand in the way of an effort on our part to come at the truth.... The master key of knowledge is, indeed, a persistent and frequent questioning."[4] Abelard even went so far as to invoke the authority of Aristotle, "the most clear-sighted of all the philosophers," claiming that he had maintained that by "doubting we come to examine, and by examining we reach the truth."[5]

As might have been anticipated from these short excerpts, some theologians, especially those who took a more intuitive or mystical view of religion, did not accept the appropriateness of applying logical analysis to questions of faith. Indeed, both Abelard's conclusions and his approach to faith were challenged by one of the most powerful religious thinkers of the time, Bernard of Clairvaux (1090–1153). Bernard appealed to the Council of Soissons to censure Abelard, and on June 3, 1140, they condemned nineteen propositions they attributed to him. The Council ordered Abelard to remain silent; the pope eventually rescinded this order; Abelard died in April 1142.

This tension between "rational" and "mystical" approaches to faith increased as translations of Aristotle's ideas and the philosopher Averroes' twelfth-century commentaries on them began to appear in Europe in the thirteenth century.[6] Theologians soon recognized the antagonism between Aristotle's natural philosophy and logic and Christian theology. An indication of this recognition is the Edict of Paris of 1210 that proscribed lecturing on Aristotelian ideals. Despite this edict, by the 1240s Aristotle's physics was being taught at the two intellectual and theological centers of the day, Oxford University and the University of Paris. By 1255 Aristotelian natural philosophy was firmly established at both of these institutions. One of the benchmark dates in the struggle between Aristotelianism and theology is March 19, 1255. On that day, the faculty of arts at the University of Paris placed Aristotle's writings at the center of the curriculum. This, in effect, split the university's faculty into two groups: a *philosophical* faculty, who found knowledge (but perhaps not truth) in the works of Aristotle and others, and a *theological* faculty, who sought only to teach truth as revealed through the Bible. This split was not a full-blown schism, because the philosophical faculty also took the Bible as the source of ultimate truth, but their willingness to consider the argumentation of Aristotelian natural philosophy did exacerbate the conflict between natural philosophy and theology.

Not all thirteenth-century theologians were concerned about Aristotelian ideas undermining their faith. Two theologians, Albertus Magnus (Albert the Great, c. 1193–1280) and Aquinas, wrote commentaries on works of Aristotle with few references to God, or God's power. According to the modern medieval scholar Edward Grant, Albertus wrote his commentary of Aristotle's *Physics* because "his Dominican brothers had implored him" to provide an account of physics so that they would, in Albertus' words, "have a complete science of nature and that from it they might be able to understand in a competent way the books of Aristotle."[7] His attitude toward the supposed antagonism between natural philosophy and theology can be inferred from his justification for commenting on *Physics*: There was not one. This is also evident in Albertus' commentary on *De Caelo*, wherein he questioned whether there could be other worlds, meaning inhabited heavenly bodies. Albertus indicated that God could produce more worlds, if he were so inclined, but then argued that nature itself, through natural processes, could not.

Aquinas attempted to mend the antagonism between philosophy and religion by asserting that God had revealed both supernatural and philosophical truths. As an example, Aquinas said that God revealed his existence to us, so there is some knowledge that we obtain directly from God. On the other hand, it is possible to obtain knowledge through logical argumentation; for example, it is possible to philosophically conclude that God exists—Aquinas gave five such *proofs*. Thus truth could indeed be found in strictly philosophical texts.

The philosophical movement can best be understood through an examination of the ideas of Siger of Brabant (1240–74), a member of the faculty of arts during a tumultuous period. While the theologians sought to uncover concealed truths, Siger annunciated a principle that guided the philosophers: "It should be noted by those who undertake to comment upon the books [of Aristotle] that his opinion is not to be concealed, even though it be contrary to the truth."[8] Thus the philosophers saw a value in examining ideas even if they conflicted with presumed theological truths.

For more conservative theologians, even Aquinas went too far toward admitting that knowledge, and so truth, could be found outside faith. They saw that the acceptance of either Siger's or Aquinas' views would force theologians into endless attempts to reconcile conflicts between these two possible sources of truth, all the while recognizing the authority of the Bible. Perhaps they even sensed that some philosophical explanations might prove to be more satisfying than appeals to faith. One reaction to Aquinas' accommodation of philosophical ideas came in 1272, when the faculty of arts in Paris decreed that no faculty member should presume to dispute any purely theological matter. The motion the faculty passed also addressed the question as to whether there could be two paths to truth: "If any [faculty member] reads or disputes any difficult passages or any questions which seem to undermine the faith, he shall refute the arguments or texts as far as they are against the faith or concede that they are absolutely false."[9]

This decree undoubtedly left many issues unresolved, and it offered no guidance on how to address them. Scholastics were asking questions upon which the Bible had no opinion, yet felt it necessary to resolve them in a manner consistent with biblical interpretation. This precarious balancing act was overturned a few years later when Bishop Etienne Tempier of Paris issued the infamous condemnation of 1277. This condemnation listed 219 errors of thinking, or belief, for which someone could be excommunicated. Two errors listed in the condemnation addressed the split among the faculty at the university:

Error 40. There is no higher life than philosophical life.
Error 154. [The] only wise men of the world are philosophers.[10]

Besides anointing the theologians as the only thinkers with access to truth, the condemnation of 1277 resolved the fundamental question of whether God's power enables him to do things contrary to principles of Aristotelian natural philosophy. God's power was absolute.

Despite the condemnation of 1277, discussions of Aristotelian principles continued throughout the Middle Ages, and sometimes came dangerously close to heresy. And although theology was not as axiomatized as Boethius might have hoped for, Scholastics frequently appealed to the principles of Aristotelian logic to support their conclusions. The twentieth-century philosopher John Carnes has written on the parallels between theology and axiomatic systems. In his book *Axiomatics and Dogmatics* Carnes compares theology with science, at least science as it is imagined to proceed according to one widely accepted model. This is the so-called received view of science, which according to Carnes, holds that a scientific theory

is formulated in terms of an axiomatized, logical structure;
consists of language that embodies logical terms, observational
    terms standing for observed or observable phenomena, and
    theoretical terms peculiar to the theory under consideration; and
provides correspondence rules connecting the theoretical
    vocabulary with observational vocabulary.[11]

Any axiomatized, logical structure includes what Carnes called *primitive*, or undefined, terms; a set of unproven but presumably self-evident, true statements; and a set of rules for inferring new true statements from the axioms. Both Euclid and Anselm attempted to provide definitions for their most fundamental terms, for Euclid *a point* and for Anselm *God*. But saying that a point is *that which has no part* does not tell you what it is, just as saying *God is that than which no greater can be conceived*, only tells you what God is not. Neither of these expressions is a definition, each is a property of an entity whose existence is already assumed and whose essence is understood. Other important primitive elements in medieval theology are souls, heaven, and angels. Of these, angels most deserve our attention.

QUESTIONS ABOUT ANGELS

Of all the questions you might want to ask
about angels, the only one you ever hear
is how many can dance on the head of a pin.

No curiosity about how they pass the eternal time
besides circling the Throne chanting in Latin
or delivering a crust of bread to a hermit on earth
or guiding a boy and girl across a rickety wooden bridge.

Do they fly through God's body and come out singing?
Do they swing like children from the hinges
of the spirit world saying their names backwards and forwards?
Do they sit alone in little gardens changing colors?

— *Billy Collins, "Questions about Angels" (1999)*

The contemporary American poet Billy Collins speculates on the types of questions a modern person might ask about the nature of angels and juxtaposes them with the apocryphal one of how many could dance on the head of a pin:

> the medieval theologians control the court.
> The only question you ever hear is about
> the little dance floor on the head of a pin.[12]

Although there is no evidence that any notable theologian or philosopher ever attempted to provide an estimate for the number of dancing cherubs confined to a small space, nor that anyone ever imagined trying, this question is typical of the sorts of questions that occupied Western thought in the late Middle Ages and early Renaissance.

What is striking about such questions is their naive intertwining of physical theories, mathematical assumptions, and metaphysical beliefs. For example, the question of

whether angelic power could produce a vacuum in nature

juxtaposes one assumption of medieval Christian thought, that angels exist, and can act, with an assumption of Aristotelian natural philosophy, that a vacuum cannot exist in nature. It is evident in the very posing of this question that medieval Scholastics struggled with conflicts between their theology, the concept of an all-powerful God, and their understanding of, for lack of a better word, physics.

Edward Grant has cataloged many of the questions the Scholastics did address. These appear as an appendix in his book *Planets, Stars, and Orbs*. Many of these queries deal indirectly with the conflict between Christian and Aristotelian philosophies, such as the question about the power of angels to produce a vacuum (question 322 in Grant's list) and the question

whether there is something beyond the sky. (question 54)

Others are relatively straightforward questions about the material world:

Whether the nonstarry part of the heaven is visible to us.
(question 136)
Whether celestial bodies cause sound by their motions.
(question 222)
Whether it is possible that a vacuum can exist naturally.
(question 308)

The aspect of the Scholastic attempts to answer these questions that is most relevant to the ideas of this book is their reliance on deductive reasoning. Proof was their purported method for uncovering truth. The Scholastics based all of their reasoning on three central axioms—the unerring truth of Christian theology, the essential correctness of Aristotelian natural philosophy, and the soundness of mathematics. Working from these axioms, the Scholastics attempted to reconcile both observed and imagined conflicts between the theological and philosophical cosmologies. These arguments forced them into examinations of their understanding of mathematics, infinity, beauty, and truth, and because religious faith was the ultimate source of truth, changed the dominant conception of each.

WHETHER THE EMPYREAN HEAVEN IS A BODY (QUESTION 107)

By the end of the Middle Ages, the generally accepted cosmology was a theological adaptation of the one developed by Ptolemy in the second century A.D. The structure of this theological cosmos was described in Dante's *Divine Comedy* (early fourteenth century); Dante's cosmos retains the basic structure of Ptolemy's, but it is not differentiated into just the terrestrial and celestial realms. Theological regions dominate this universe. Beneath our feet, inside the spherical earth, lay the delineated circles of hell; on the portion of earth opposite Jerusalem are the levels of purgatory. Heaven begins with the moon and continues beyond to a fixed sphere of the stars to a crystalline sphere, and depending on the answer to question 107, perhaps to a material paradise beyond the stars. This is the immobile empyrean paradise.

In *The Divine Comedy*, Beatrice takes the narrator, presumably Dante, on a tour of the cosmos. After Beatrice educates Dante about the nature of hell and purgatory, in books 1 and 2, book 3, *Paradise*, begins with their ascent into the heavens. Both hell and purgatory are connected to earth, so there was no physical impediment to Dante and Beatrice touring them, but heaven is distinct from earth. Dante is concerned with how he is to ascend into the heavens; Beatrice's explanation reveals an important difference between Ptolemy's heavens and the one presented by Dante:

> Among themselves all things
> Have order; and from hence the form, which makes
> The universe resemble God. In this
> The higher creatures see the printed steps
> Of that eternal worth, which is the end
> Whither the line is drawn. All natures lean,
> In this their order, diversely, some more,

Some less approaching to their primal source.
Thus they to different havens are moved on
Through the vast sea of being, and each one
With instinct given, that bears it in its course.[13]

So not only are there souls in hell and purgatory, all heavenly regions are occupied. The planetary spheres are not lifeless, geometrically driven objects, but the realm of both human souls and angels. The closer a sphere to the earth, that is the farther from the empyrean paradise, the less perfect the souls or angels who reside there. For example, in Dante's first heaven, the sphere of the moon, reside both the souls of people who broke their vows and the guardian angels of individuals. Theologians and the angels who combat evil occupy the fourth heaven, the sphere of the sun. The crystalline sphere is the last sphere revolving around the earth, and therein reside the angels that revel in God's divine love. The empyrean paradise is reserved for only the purest souls, those who are worthy of the complete understanding that comes from proximity to God. To be in the empyrean paradise an individual must indeed have been devout, such as the Virgin Mary. Dante also places Bernard of Clairvaux, now Saint Bernard, in the empyrean even though Bernard had led the failed Second Crusade (1147–49) in which Dante's great-great-grandfather was killed.

In Dante's universe it is not only the heavens that are organized along geometric lines. Just as the spheres of the planets differentiate the sky into separate regions, hell and purgatory are partitioned into levels. In the *Divine Comedy*, Dante systematically describes the delineations of hell, purgatory, and heaven.

HELL

If anyone says or thinks that the punishment of demons and of impious men is only temporary, and will one day have an end, and that a restoration will take place of demons and of impious men, let them be anathema.
— *The Seven Ecumenical Councils of the Undivided Church*, "Decree of Synod of Constantinople" (553)

The hell described in Milton's *Paradise Lost* is a world where "Night / And Chaos, ancestors of Nature, hold / Eternal anarchy, amidst the noise

/ Of endless wars, and by confusion stand." Although chaos remains the dominant attribute of Dante's hell, in the *Divine Comedy* there is an order to the degrees of chaos. Hell is visualized as a deep pit with different levels; these are circular cliffs located at successively lower depths along the pit's perimeter. These levels are known as the nine circles of hell. The first circle is not hell proper but limbo. This is a place where virtuous pagans and unbaptized children spend their eternity. These souls are not subjected to any physical torment, but instead they suffer only through being removed from God's glory. In *The Inferno*, Dante lists some of limbo's inhabitants: Dante sees Aristotle "amid the philosophic train. / Him all admire, all pay him reverence due." Dante continues:

> There Socrates and Plato both I marked,
> Nearest to him in rank; Democritus,
> Who sets the world at chance, Diogenes,
> With Heraclitus, and Empedocles,
> And Anaxagoras, and Thales sage.[14]

The second circle of hell is the first level of the fire-and-brimstone hell of modern fundamentalism. Minos, the mythological king of Crete who was known for his wise judgment, sits here, awaiting the arrival of the souls to which he assigns a torment particular to their sin. Minos sends poets back to the first circle, retains the carnal—those who gave in to excessive desire—and sends the more sinful to an appropriate lower level with a specialized torment. The ninth circle of hell is not a circle but the bottom of the pit. It is the home of Satan and is reserved for only the most sinful.

As Dante travels with Beatrice he describes the different circles of hell and their inhabitants, which are summarized below.

CIRCLE OF HELL	INHABITANT'S SIN	POSSIBLE TORMENT
first	paganism	hopelessness
second	lust	whirled by murky wind
third	gluttony	half buried in garbage
fourth	hoarding, wasting	eternal conflict
fifth	wrath, sullenness	war in swamp
sixth	heresy	flames

seventh	violence	immersed in boiling blood
eighth	fraud, flattery	hung by their heels and burned
ninth	treachery	frozen in ice

Dante's purgatory is also organized into levels. Many of the souls in purgatory have given into the same vices as the souls of hell, but these are the souls of the saved. Depending on its vice, a soul is assigned to one of seven levels and is subjected to a sort of deprogramming. Once a soul has been purged of its vice it ascends to heaven.

WHETHER THE SPOTS APPEARING IN THE MOON ARISE FROM DIFFERENCES IN PARTS OF THE MOON OR FROM SOMETHING EXTERNAL (QUESTION 141)

This to the lunar sphere directs the fire;
This moves the hearts of mortal animals;
This the brute earth together knits, and binds.
Nor only creatures, void of intellect,
Are aimed at by this bow; but even those,
That have intelligence and love, are pierced."
—Dante, Paradise *(14th century)*

Despite the complexity of Dante's cosmos, the universe, like Aristotle's and Ptolemy's, is essentially divided into two realms, the tangible terrestrial world and the inaccessible celestial world. All heavenly bodies are nonterrestrial, and the earth and its objects are terrestrial, but that does not provide a clear boundary between the two.

Aristotle attempted to clarify the differences between heaven and earth by describing the nature of the matter each contains. In the *Divine Comedy* these differences are based on the nature of its inhabitants. The Aristotelian terrestrial world consists of all matter that is composed of the four primary elements earth, air, fire, and water—an object or body is terrestrial only if it is made up of these elements. Aristotle postulated that the objects in the celestial world were made up of some fifth, unknowable element.

But it was apparent to Aristotle, and everyone else, that some terrestrial matter does not remain earthbound; fire is inclined to move into the heavens. Aristotle did not view the rising of fire (smoke) as exceptional behavior for one of the primary elements, but as an extreme case among a range of possible behaviors—for example, water always sought to lie below air. Each of these tendencies is explained through Aristotle's theory of heavy and light. In this theory the four elements are arranged, from heaviest to lightest: earth, water, air, fire. This ordering more or less agrees with observation; if there is no interference then water lies between earth and air, and fire resides somewhere above air. However, fire's unfortunate inclination to move upward, toward and possibly into the heavens blurs the distinction between the terrestrial and heavenly worlds based on their fundamental elements. Perhaps it could have been postulated that fire transforms into celestial matter, but even this would obscure the separation of the earth from the sky. Aristotle sidestepped this problem by simply maintaining that fire does not move into the heavens but remains in the region just below the position of the moon.

Aristotle's theory of heavy and light presented him with the following difficult question: If fire is the lightest of all of the elements why is it still present on earth at all? Why haven't all of the fire-elements drifted off into the sublunar region? There are two parts to Aristotle's answer, and they both involve the distinction between terrestrial and heavenly space. The first part of Aristotle's solution is his observation that in the terrestrial world, change and decay are pervasive. Accepting Plato's transformation cycle, if not its geometric basis, Aristotle believed that fire is continually being remade from air; more fire is always available. This still does not address the problem presented by fire's continual rise toward the heavens, which brings us to the second part of Aristotle's answer. Although he did not use this terminology, the earth-air-fire-water system is a closed system. The moon's orbit provides a barrier through which fire cannot escape. Fire remains within the system to be further transformed.

Just beyond the ring of fire there is evidence that something is wrong with all of the cosmologies we have discussed; this evidence is spots on the moon, what we call "the man in the moon." Dante saw these markings during his initial ascent into the heavens and asked Beatrice, "But tell, I pray thee, whence the gloomy spots / Upon this body?"[15] These markings conflict with the assumption that all heavenly bodies are perfect spheres, and these markings provide visible evidence that the mathematically inspired mechanisms for planetary movements cannot be correct.

According to Plutarch, Pythagoras, whose sense of mathematical perfection came to dominate cosmology, accounted for the spots on the moon by declaring that the moon was terrestrial. The Pythagoreans affirm "that the *Moon* appears to us Terraneous [terrestrial], by reason it is inhabited as our Earth is, and in it there are Animals of a larger size, and Plants of a rarer beauty than our Globe affords, and that the Animals . . . are fifteen degrees superior to ours; that they omit nothing Excrementitious; and the days are fifteen times longer."[16]

Since, for Aristotle, the heavens began just beyond the realm of fire, the moon was a transitional body, neither entirely heavenly nor entirely terrestrial. Averroes claimed that Aristotle dismissed the moon's imperfection by maintaining that the moon is composed of some combination of fire and the heavenly fifth element. The spots on the moon are places where these two elements are not homogeneously mixed. Thus the moon is intermediate between heaven and earth.

Theologians and Scholastics also addressed the difficulties presented by the man in the moon. In 1627 the Italian theologian Raphael Aversa (1589–1671) surveyed nine different explanations for the lunar spots. These can be broadly characterized as:

1. The spots are illusions, depending on shadows or reflections.
2. The spots are the result of terrestrial events.
3. The spots are intrinsic to the composition of the lunar surface.[17]

Plutarch had suggested that the lunar spots could be caused by vapors interposed between the earth and the moon, but Aversa rejected explanation number 1 because the spots are unchanging and appear the same from all points on earth (at least as far as he knew). The first-century Roman encyclopedist Pliny the Elder (23–79) had concluded that terrestrial vapors rose from earth and adhered to the lunar surface, affecting the moon's luminosity or reflectivity. But Aversa believed that terrestrial vapors, or material, could never penetrate the region of the heavens, so he dismissed Pliny's explanation. This leaves the third possibility; the lunar spots are on the moon, or at least manifestations of some property of its surface. The details of this explanation of the spots depended, in part, on how the advocate answered the following:

> Whether all the planets, except the sun, receive their light from the sun or from themselves. (question 231)

Because of its phases, the moon was not considered to be strictly self-luminous, so the above question was transformed to another: Is the moon solely reflective, or does it possess some property of luminosity? This second possibility is best understood through a short quotation from Averroes: "It has been demonstrated that if the moon acquires the power of lighting up from the sun, it is not from reflection. . . . If it illuminates, it is by becoming a luminous body itself. The sun renders it luminescent first, then the light emanates from it in the same way that it emanates from the other stars; that is, an infinite multitude of rays are issued from each point of the moon."[18]

Whether the moon is entirely reflective or strangely luminescent, as explained by Averroes, we are left to explain how a perfectly uniform heavenly body could have parts with different properties. There are also two answers to this: That the variation is caused by parts of the moon being of different densities or that the moon is an uneven mixture of two or more elements. Beatrice refutes both of these in Dante's *Paradise*. In rejecting the first explanation Beatrice says:

> If rarity were of that dusk the cause,
> Which thou inquirest, either in some part
> That planet must throughout be void, nor fed
> With its own matter; or, as bodies share
> Their fat and leanness, in like manner this
> Must in its volume change the leaves. The first,
> If it were true, had through the sun's eclipse
> Been manifested, by transparency
> Of light, as through aught rare beside effused.
> But this is not.[19]

Beatrice then rejects the second possibility, by asserting that the reflection of light would be the same off of the thick and thin sections, and so the differences would not be visible to us.

Aversa reached the same conclusion as Averroes, whose explanation for the lunar spots dominated medieval thinking: The light from the sun excites the various parts of the moon, which then become luminescent, but not all portions become equally luminescent. The parts that become the least luminescent are the ones that we see as the lunar spots. This is not due to different densities but, somehow, to different portions of the moon having different opacity.

It became the accepted view that the moon is somewhere between heaven and earth and thus shares properties with both. This, in itself, presented philosophical difficulties. Central to Aristotle's cosmology, and then Christian theology, are the uniqueness of the earth and the perfection of the heavens. If the man in the moon is visible evidence that the surface of the moon shares some features with the surface of the earth, then the uniqueness of the earth, as the only world, is challenged. If, on the other hand, the man in the moon is evidence for some uneven mixing of heavenly materials, then the perfection of the heavens is challenged.

WHETHER THE HEAVENS OR PLANETS ARE MOVED BY INTELLIGENCES OR INTRINSICALLY BY A PROPER FORM OR NATURE (QUESTION 195)

The existence of the man in the moon challenged not only the assumed perfection of all heavenly bodies but also the mechanics of Ptolemy's system of epicycles. This is because Ptolemy's model, although using circular orbits in a fundamentally different manner from Aristotle's, is still based on Aristotle's celestial physics. A central tenet of Aristotelian physics is that neither the moon nor any other heavenly body can rotate about its own axis. It is one of those curious pieces of intellectual history that Aristotle used one observation that is contrary to the underlying principles of his cosmology, the existence of the man in the moon, to support his position that no heavenly body can rotate on its own axis.

Aristotle argued that since the moon moves around the earth, as described in his model, then the moon could not rotate because the man in the moon always faces the earth. If the moon rotated as it revolved around the earth, the man in the moon could not always aim inward. Thus the moon must move as if it is firmly attached to the circle of its orbit.

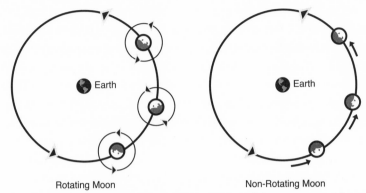

Rotating Moon   Non-Rotating Moon

FIGURE 5.1. These two drawings illustrate the movement of the moon around the earth in Aristotle's model. In the drawing on the left, if the moon were to rotate on its axis as it revolves around the earth, then the man in the moon would not always face the earth. The drawing on the right illustrates that if the moon does not rotate on its axis, a rotation Aristotle's physics prohibits, then the man in the moon always faces the earth.

However, if we adopt both Ptolemy's model of epicycles and the Aristotelian principle that the moon cannot rotate about a central axis, then whichever side of the moon faces the center of the epicycle will always face the center of the epicycle. This means that the lunar spots will sometimes face away from earth, which could only be resolved in Ptolemy's model by allowing the moon to rotate as indicated by the sequence in Figure 5.2.

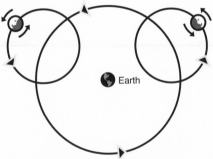

FIGURE 5.2. Ptolemy's use of epicycles requires that the moon rotate on its axis in order for the man in the moon to always face the earth.

This rotation of the moon undermines the universal validity of Aristotelian physics. Any theologian or philosopher attempting to answer question 195 (above), which was addressed by Aquinas among others, would have to appeal to the presumed properties of angels, and these would have to be either self-evident or previously deduced. According to Grant, "The properties attributed to [angels] were usually those that Aristotle had attributed to the prime mover . . . immobility, indivisibility, and absence of magnitude. Although angels possessed considerable power, they could not exceed the finite powers conferred on them at the Creation."[20] Those who worried about such matters sought to explain the different intensities of the stars, the mechanism for the motion of epicycles, and the troubling existence of the man in the moon. One fail-safe solution was to invoke God's will, usually as carried out by angels.

For example, in the following lines from *Paradise*, Beatrice offers the standard explanation for the different intensities of the stars in the celestial sphere and why they twinkle—these differences are caused by the differences among the angels (intelligences) residing in this eighth sphere. The light from each star shines through an angel's body, and as the angels have different powers, they diffuse the light to different degrees.

> the soul, that dwells within your dust,
> Through members different, yet together formed,
> In different powers resolves itself; e'en so
> The intellectual efficacy unfolds
> Its goodness multiplied throughout the stars;
> On its own unity revolving still.
> Different virtue compact different
> Makes with the precious body it enlivens,
> With which it knits, as life in you is knit.
> From its original nature full of joy,
> The virtue mingled through the body shines,
> As joy through pupil of the living eye.
> From hence proceeds that which from light to light
> Seems different, and not from dense or rare.
> This is the formal cause, that generates
> Proportioned to its power, the dusk or clear.[21]

Beatrice also explains that the moon appears to have spots because of the differences in the powers of the angels that occupy its sphere.

Epicycles were perhaps more troubling because geometry was the preferred mechanism for planetary motion. Aristotle had attributed all initial cosmic motion to an unmoved mover, and although he described celestial motion by appealing to a system of gears moving gears, this is not the image Aristotle presented in his *Metaphysics*. There he attributed to each planetary orb an "immovable mover," an external, perhaps spiritual, mover responsible for its movement. The mover of the celestial sphere is the "first immovable mover," what became known as the primary or unmoved mover.

In Dante's cosmos the unmoved mover is God, and the movements of epicycles are explained by the continuing intervention of God in the workings of the universe. God does not cause the heavenly motions directly, but through the angels associated with each body. For example, as Dante and Beatrice approach the sphere of Venus, having just visited Mercury, Dante says:

> To those celestial lights, that towards us came,
> Leaving the circuit of their joyous ring,
> Conducted by the lofty seraphim.[22]

These angels then begin to sing to Dante and Beatrice, telling them of their role in the workings of the cosmos. The angels sing that they are the ones

> To whom thou in the world erewhile didst sing;
> "O ye! whose intellectual ministry
> Moves the third heaven": and in one orb we roll,
> One motion, one impulse, with those who rule
> Princedoms in heaven; yet are of love so full,
> That to please thee 'twill be as sweet to rest.[23]

Medieval Scholastics debated whether these angels were external or internal movers, whether they were different from or intrinsic to the heavenly spheres, and whether they had power independent of God. In each of these cases, angels appeared to have the ability to overrule the principles of Aristotelian physics and do the impossible. Divine intervention in heavenly motions freed Dante from the rules of Aristotelian physics; the laws for heavenly motion were ultimately metaphysical, thus faith based.

POSTSCRIPT: THE SIZE OF THE COSMOS:
INFINITY RECONSIDERED

It has been proved that the distance between the centre of the earth and the outer surface of the sphere of Saturn is a journey of nearly eight thousand seven hundred solar years. Suppose a day's journey to be forty legal miles ... and consider the great and enormous distance!

— *Maimonides,* Guide for the Perplexed *(12th century)*

The existence of epicycles, which were thought to be material objects and so could not overlap, forced Ptolemy to reconsider earlier estimates for the size of the universe. In Ptolemy's cosmos, some epicycles were fairly large, certainly larger than the objects they moved. To fit all of these epicycles into the heavens, Ptolemy had to posit a larger universe. Also, since there are no gaps between the Ptolemaic spheres of the heavenly bodies, as there cannot be a vacuum in space, the moon's greatest distance from the earth had to equal Mercury's least distance; Mercury's greatest distance had to equal Venus' least distance, and so on. There are two distances for each body since, in Ptolemy's model, every heavenly body moves around the earth on an epicycle, not along a perfectly circular orbit, and so has a closest and farthest distance from earth. The unit of measurement e.r. is the radius of the earth.

HEAVENLY BODY	CLOSEST DISTANCE FROM EARTH (in e.r.s)	GREATEST DISTANCE FROM EARTH (in e.r.s)
moon	33	64
Mercury	64	166
Venus	166	1,079
sun	1,160	1,260
Mars	1,260	8,820
Jupiter	8,820	14,187
Saturn	14,187	19,865
fixed stars	20,000	20,000

There are two "gaps" in this table: between the sphere of Venus and the sphere of the sun and between the sphere of Saturn and the fixed stars. According to Albert van Helden, Ptolemy obtained the first gap by using the nesting of spheres to find the greatest distance to Venus and the second gap by using Aristarchus' eclipse diagram to find the least distance to the sun. Ptolemy knew he could have used his data to close up the gap between Venus and the sun, but not the gap between Saturn and the fixed stars.[24]

The Arab astronomers of the late first millennium knew of Ptolemy's *Almagest* several hundred years before its transmission to Europe. They made improvements to Ptolemy's estimates, and in their work they converted the unit one earth radius into miles. They estimated that 1 degree around the earth was $56\frac{2}{3}$ miles; hence the circumference of the earth was calculated to be $360 \times 56\frac{2}{3} = 20,400$ miles. Thus it was generally accepted that the sun was, on average, $3,926,450 = 1210 \times 3,245$ miles away (the modern estimate is approximately 93,000,000 miles, so 23 to 24 times farther). More significantly, the fixed stars were imagined to be $64,900,000,000 (= 20,000 \times 3,245)$ miles from earth.

Archimedes, and then others, devised schemes for representing large numbers; for example in the table above these distances were recorded in terms of the radius of the earth, e.r.s, instead of in miles, but these numbers were still incomprehensible. Some writers attempted to translate these numbers into a more understandable language. For example, in *The Guide for the Perplexed*, Maimonides wrote, "In order to obtain a correct estimate of ourselves, we must reflect on the results of the investigations which have been made into the dimensions and the distances of the spheres and the stars." He then went on with the estimate for the distance from earth to Saturn that introduced this section (a distance that would take a walking person 8,700 years to traverse). Maimonides then argued that the celestial sphere must be very thick, since the "body of each of these stars is more than ninety times as big as the globe of the earth," further adding to the size of the universe.[25]

In the thirteenth century, Goussouin de Metz composed a poem "Image du Monde" (1246), in which the Arab astronomer al-Farghani's ninth-century version of Ptolemy's *Almagest* is taken as fact. Goussouin calculated how long it would take Adam to walk from the earth to the sphere of the fixed stars had he started at the moment of Creation and walked twenty-five miles per day. Taking the year of Adam's creation as $5199\frac{1}{2}$ B.C., Goussouin calculated that Adam would have reached the stars in 1958.

When Copernicus moved the sun to the center of all planetary orbits, both the size and structure of the cosmos were forever changed. A heliocentric universe was not new with Copernicus; Aristarchus had already suggested this. Aristarchus had envisioned precisely the same arrangement as Copernicus: the planets, including the earth, orbit the sun, and the moon orbits the earth. To account for the motion of the stars, Aristarchus' earth rotated on an axis. But by the third century A.D., the principles of Aristotelian physics were already firmly entrenched, and because of Aristotle's theory of place, the concept of a moving earth was especially unacceptable. After all, it was argued, if the earth were moving an object would lag behind when dropped, thus moving along a curve as it fell to the ground; the clouds would trail off into the heavens.

Copernicus knew the shortcomings of Ptolemy's view of the cosmos, and of the continued conflict between that model and the principles of Aristotle's physics. He experimented with Aristarchus' heliocentric universe and decided that putting the sun in the center of the universe corrected some of the errors of Ptolemy's system. Copernicus did not give a reason for why the entire cosmos revolved around a stationary sun, he knew any such speculation would put him into a direct confrontation with the then-current theology. And because he still thought planetary motion was uniform and circular, Copernicus was forced to retain some of the spheres-revolving-around-spheres aspects of Ptolemy's model of epicycles. Having the planets revolve around the sun allowed Copernicus to model all of the planets on a simpler system of epicycles than in Ptolemy's model. By the sixteenth century the Ptolemy-based model required seventy-seven spheres in order to account for the increasingly accurate celestial data. Copernicus' model required only thirty-four spheres.

Philosophically the Copernican model was radical, not just because it moved the earth away from the center of the universe, but also because it expanded the size of the universe. Since the earth is now moving around the sun, it is closer to different portions of the sphere of stars at different times during its orbit. This means, as the criticism of the Copernican model went, that we should be able to discern changes in the brightness of any particular star as the earth orbits the sun. To meet this objection, Copernicus correctly concluded that the sphere of stars is so distant that any variation in the size, and hence brightness, of a star is imperceptible.

Moving the sun near the center of all planetary movements forced Copernicus to recalculate the distances of the planets from the earth, and at first glance these calculations appear to be in conflict with Copernicus' larger universe. What they show is that the solar system is smaller than Ptolemy had believed. Saturn is significantly closer to earth, but the distance from Saturn to the backdrop of fixed stars is four hundred times as far. Thus in the sixteenth century the medieval estimate for the size of the universe, which was just under 65 million miles, was replaced by one of more than 25 billion miles.

In one his poems, "Considering the Void," former president Jimmy Carter expresses the same distress others have felt while contemplating cosmic distances and the multitude of stars in the universe. In Ptolemy's cosmology the number of stars was a comfortable 1,022. In his poem Carter refers to "an infinity of Suns," but allows that the number might only be a "thousand billion" (i.e., 1,000,000,000,000).[26] Perhaps even more disturbing is the thought that the universe, and possibly the number of stars in it, might not be a measurable quantity. The philosopher Martin Buber (1878–1965) attributed his contemplation of infinite space or time with challenging his sanity: "A necessity I could not understand swept over me: I had to try again and again to imagine the edge of space, or its edgelessness, time with a beginning and an end or time without beginning or end, and both were equally impossible, equally hopeless. . . . Under an irresistible compulsion I reeled from one to the other, at times so closely threatened with the danger of madness that I seriously thought of avoiding it by suicide."[27]

Although very few ancient philosophers contemplated an infinite cosmos, some did. For example, Democritus and Lucretius both argued that the universe must be infinite. By the Middle Ages some writers saw the Aristotelian prohibition on the existence of a material infinitude to be at odds with the existence of a metaphysically infinite, absolutely powerful God. This prohibition seemed to limit both God's material (or physical) scope and God's ability to create. The Scholastics who saw this conflict were compelled to either resolve or refute it.

The dominant theological view in the Middle Ages was that God's creative powers did not end with Genesis, so Aristotle's finite, complete space, although it was consistent with theology, appeared to limit God. In the thirteenth century Walter Burley (c. 1275–1344) examined this conflict and concluded that God's creative ability had no limit and yet Aristotle's rejection of infinite magnitudes was correct.[28] Burley proposed that the belief that God could create a magnitude, or quantity, greater than any specified one, does not necessarily mean that God can create an existing, infinite quantity. For example: Could God create an endless sequence of boxes within boxes; that is, starting with a single small box can God create a slightly larger box containing that first box, then another box containing the first two nested boxes, and so obtain an endless sequence of boxes within boxes? Burley concluded that no matter how many nested boxes have been created, God could create another box containing the original collection of boxes. However, God cannot create an actual existing infinite chain of boxes within boxes. Burley thus agreed with one Aristotelian principle, that there cannot exist an infinite entity, but disagreed with another, that neither matter nor space can be created nor destroyed. Burley's God had created a finite universe that is potentially infinite by addition.

In the thirteenth century, Aquinas took up the question of whether God could create an infinite magnitude. Indeed, Aquinas offered the entirely sensible opinion that not being able to do the physically impossible is no limitation on God's power: "If the infinite can exist in actuality, according to the nature of things . . . I state that God can create an actual infinity. But if actual existence is repugnant to infinity due to its own ground, then God would not be able to produce this existence, no more than He would be able to make it that man were not a rational animal."[29]

Gregory of Rimini (c. 1300–1358) understood that Burley's reasoning could be amended to prove that God can indeed create an existing infinite quantity by importing the concept of time underlying Zeno's Dichotomy paradox into God's creative processes. A central assumption is that God's omnipotence means that he can create an entity, for example an angel, in as short an interval of time as desired. Gregory appealed to this process to demonstrate that God could produce an infinite collection of angels. Suppose an arrow is shot at a target (and contrary to Zeno's paradox the arrow does begin to move). While the arrow moves from the bow to halfway to the target, God creates an angel.

Then, while the arrow moves from that position to halfway to the target, God produces another angel. As there are an unlimited number of these increasingly short time intervals, by the time the arrow hits the target God will have produced an unlimited number of angels. (This argument describes how an omnipotent God could create an infinite number of angels; it can be modified to demonstrate that an omnipotent God could create both an infinite collection and an infinitely large physical object. Instead of creating an angel at each step in the process, as the arrow moves from the bow to its target, God could create a cubic foot of stone. Thus God will have created both an infinite quantity, the number of blocks of stone, and an infinitely large material object, for example a tower built by piling these stones one atop the other.) Gregory's point was that it is logically possible to have a mathematically infinite quantity, and therefore there is nothing inherent in the natural world to prohibit it.

There were philosophical reasons for imagining that God had to have created an infinite cosmos. Part of the dominant view of God was derived from Plotinus, who had endowed the metaphysically infinite One with benevolent attributes. Plotinus also gave voice to a concept concerning the natural world that combines ideas from Plato, *plenitude*, and from Aristotle, *continuity*, that eventually led some philosophers and Scholastics to conclude that the universe must be infinite. According to Lovejoy, Plato put forth in the *Timaeus* the idea that in the world, "the range of conceivable diversity of *kinds* of living things is exhaustively exemplified."[30] This idea was extended to the concept of plenitude that Lovejoy defined as the belief that "the extent and abundance of the creation must be as great as the possibility of existence and commensurate with the productive capacity of a 'perfect' and inexhaustible Source."[31]

Although Aristotle did not agree with this principle, he did contribute the second idea to Plotinus' attribute of One—*continuity*. According to Aristotle, nature "passes so gradually from the inanimate to the animate that their continuity renders the boundary between them indistinguishable."[32] Plotinus combined plenitude with continuity and articulated the existence of what became known as the *chain of being*, the idea that the universe is "composed of an immense . . . number of links ranging in hierarchical order from the meagerest kind of existents

...up to ... the Absolute Being."[33] Thus the stage was set for the debates concerning whether or not God could have, or more to the point, should have, created only a finite universe, and what is the position of humans in the hierarchical order of beings.

In the fifteenth century, Nicolas de Cusa (1401–64) argued that because God is infinitely powerful, he would have created an infinite cosmos containing infinitely many stars. Nicolas de Cusa further argued that each of these stars must necessarily have planets (*worlds*) orbiting them, and that these worlds must necessarily be inhabited. One hundred fifty years later, Giordano Bruno (1548–1600) appealed to plenitude to support his belief that the universe is infinite, and contains infinitely many worlds:

> I do not insist on infinite space, nor is Nature endowed with infinite space for the exaltation of size or of corporeal extent, but rather for the exaltation of corporeal natures and species, because infinite perfection is far better presented in innumerable individuals than in those which are numbered and finite. . . . But since innumerable grades of perfection must, through corporeal mode, unfold the divine incorporeal perfection, therefore there must be innumerable individuals, those great animals, whereof one is our earth, . . . and to contain these innumerable bodies there is needed an infinite space.[34]

For espousing a view contrary to the dominant one, Bruno was burned at the stake in February 1600.

6 Time, Infinity, and Incommensurability

From these unequal motions of the planets, mathematicians have called that the great year, in which the sun, moon, and five wandering stars, having finished their revolutions, are found in their original situation.

— Cicero, On the Nature of Gods *(45 B.C.)*

Before the development of the clock we sensed our daily lives not as a continuous series of moments, one following the other as we moved from the past to the future, but as disconnected episodes not adhering to any discernable pattern or logic. Our world was punctuated by terrestrial events, the increasing and decreasing turbulent flow of rivers and streams, the development then dissipation of clouds, the unexplainable earthquakes that occasionally reshaped the landscape. The only observable physical regularities resided in the heavens; the movements of heavenly bodies provided a background against which the longer cycles of our lives could be gauged. There was no notion of independently measuring time to understand the celestial cycles; the heavenly cycles provided the only reliable standard.

These astronomical cycles could be taken as curiosities; every so often the planets line up in a special way, or Mars passes behind a full moon. But the regularity of these cycles, the observation that everything in the sky moves according to some predictable, repeating cycle, has given them special meaning. For example, the Pythagoreans would have taken the perceived regularity of heavenly motions as evidence for their view that mathematical/musical harmonies guide all physical processes. Others have interpreted these cycles metaphysically: The movements of the planets through the major constellations are the basis for astrology, both the ancient Chaldean form, where the skies foretold fortunes and events, and the modern form, where the skies tell us whether or not we are going to have a good day. Thus these small cycles in time were imagined to determine either individual or cultural fates.

Plato saw a different sort of meaning in the cyclic patterns in the sky; he thought that the regularity of celestial motions not only measured time but also was inseparable from its existence. For Plato, the heavenly movements did not just provide us with a timepiece; they generated time. And owing to the cyclic nature of the movements, Plato concluded that time itself must be cyclic: "Only a very few men are aware of the periods of [the planets] . . . so bewildering are they in number and so amazing in intricacy. . . . None the less it is perfectly possible to perceive that the perfect temporal number and the perfect year are complete when all eight orbits have reached their total of revolutions relative to each other."[1] The eight orbits Plato refers to are those of the five visible planets, the sun, the moon, and the celestial sphere.

In the quote that opens this chapter, Cicero refers to mathematicians who called Plato's *perfect year* the "great year." Mathematicians are involved in these speculations because any two *commensurable* cycles can be combined mathematically to form a longer, common cycle. And in the material world the Pythagorean commensurability assumption— given any two mathematical lengths we can find a ruler that allows us to simultaneously measure these two lengths with whole numbers— holds because all measurements are rational. In our measurement of heavenly cycles, the *ruler* we are using is sun-cycles, or days; if the cycle of some planet is some number of days plus a little bit, then that little bit will always be a fraction. It follows from this that any two heavenly cycles may be combined to form a longer cycle whose length is a whole number of days. This process works for any number of cycles; we can always find a larger number that they will all divide evenly. This means that if we knew the cycles of the sun, moon, planets, and stars we could find a night in the future when the sky will look exactly like it did last night. More to the point, just believing that celestial bodies follow regular cycles, without having any idea how long the cycles are, tells us that someday everything in the sky will be in precisely the same position as today, or as any day in the distant past.

The lengths of each of these grand, cosmic cycles depend entirely on two things: Which cycles are thought to be important and the perceived lengths of these individual cycles. The sun's cycle is most easily discerned. It is not that the sun's cycle is the shortest of all of the celestial cycles; such a conception of the length of its cycle would require an independent means of measurement. It is that the regularity of the sun's repeating movement from sunrise to sunset resonates with our diurnal pattern; it provides a rhythm against which both our lives and other heavenly cycles could be most easily understood. Thus the sun's cycle, one day, is not so much an independent cycle as the pulse of the cosmos.

Of the other heavenly cycles, the one most easily measured is the moon's. For one thing the moon follows an unmistakable, repeating pattern. Its phases are orderly: a thin crescent grows to a full moon, then wanes to an absence from the sky; the moon reappears as a crescent curled opposite the earlier one. Over a few nights the crescent fills in, it grows to be a quarter of the disc, then half, and eventually to another full moon. It is inconceivable that anyone could not notice at least the general pattern, that the full moon reoccurs every so often, and in between, the moon appears to have different shapes, and more importantly, that this pattern has a regularity when measured against the sun's beat. If we record, or remember, the nights in which there are full moons, we will notice that they occur every twenty-nine or thirty days. The average of the lengths of these periods is roughly twenty-nine and a half days.

This sun-moon relationship is the most easily observed connection between celestial bodies, both because the sun and moon dominate their skies and because, in human terms, their cycles are relatively short. By watching the night sky over longer periods of time, and being patient, other connections can be made. The next most distinguished celestial objects are the five planets that are visible to the unaided eye, and the most visible of these is Venus. It is usually in the sky, hovering above the horizon at either dawn or dusk.

Venus appears as a morning star for many days, disappears for a while, then reappears as an evening star before disappearing; Venus then reappears in the morning sky. (Pythagoras is credited with being the first person to realize that the morning star and evening star are the same planet.) Venus is in the early morning sky for roughly 263 consecutive days; Venus then disappears. For an average of 50 days Venus is invisible to us; it then reappears early one evening on the western horizon. Venus is in the evening sky, making each evening's first appearance in a different place, for roughly 263 consecutive days. Then, suddenly Venus is gone, invisible for 8 days, after which it reappears as a morning star. This entire cycle, from Venus's first appearance in the morning sky to its next first appearance in the morning sky, takes around 584 days.

Suppose, for example, Venus has its first appearance as a morning star on the longest day of the year, the summer solstice. Venus will have its next first appearance as a morning star 584 days later, then again 584 days after that, and so forth. From the first morning-star appearance on the summer solstice, the number of days until Venus makes its next few first morning-star appearances are

584, 1,168, 1,752, 2,336, 2,920, 3,504, 4,088, and so forth.

This cycle repeats indefinitely. What about the summer solstice? The solar year is approximately 365 days long so the number of days until the next summer solstices are

365, 730, 1,095, 1,460, 1,825, 2,190, 2,555,
2,920, 3,285, 3,650, and so forth.

If we compare the Venus-cycle and solstice-cycle lists, we see that they have a number in common, 2,920. This means that Venus will make its next first-appearance as a morning star on the summer solstice after 2,920 days. If we were to continue these lists, counting the days into the future when Venus will make its first early-morning appearance and when the sun will achieve its northernmost point in the sky, we would find other numbers on both lists such as 5,840 and 8,760. These last two numbers are simply whole number multiples of 2,920.

Plato did not give an estimate for the length of a perfect year, but first-century A.D. Greek astronomers put it at thirty-six thousand years; it has also been given as forty-nine thousand years. This notion of a great year, which is coincidental with viewing time as being cyclic, was not limited to the Greeks or to the Aegean. The Mayans called their great year "the Long Count"; it is a period after which time begins again. But the Mayans took a different cycle of Venus into account when determining their Long Count. They noticed that after completing one of its 584-day cycles, Venus does indeed start another 584-day cycle but the next one is noticeably different. Venus is still in the morning sky for 263 days, but its position along the horizon and the shape of its path through the sky are different. The Mayans also noticed that there is a pattern to the different 263-day paths Venus will follow in the morning sky. There are five distinct paths and Venus always follows them in the same order.

According to contemporary astroanthropologist Anthony Aveni, the Mayans thought of this entire sequence of five paths as Venus' cycle, not as five of the shorter 584-day cycles.[2] Thus, to the discerning Mayan eye, Venus' full cycle was $5 \times 584 = 2,920$ days long. To account for Venus' role in determining the length of the Long Count, the Mayans would have used 2,920 days instead of 584. (Coincidentally, if the Greeks had used this longer Mayan estimate for Venus' cycle it would not have made their great year any longer. This is because the length of the Greek great year was also a multiple of 2,920.) Moreover, the Mayans determined the length of their Long Count by taking into account cycles the Greeks did not acknowledge. In Mayan cosmology, there are thirteen cosmological spheres, and twenty named days. The Long Count was measured by combining two cycles: The cycle of Venus, which they took to be 2,920 days long and their 260-day ritual cycle (based on their twenty named days and thirteen cosmological spheres, $260 = 20 \times 13$). So the Mayan long count was $2,920 \times 260 = 759,200$ days, or 2,080 years long.

REFUTING CYCLIC TIME BY EMBRACING IRRATIONALITY

Whatever is imagined to happen when the celestial bodies return to some previous configuration, the belief that this happens imposes a cyclic structure on time that violates the most fundamental principles of Christian faith—the uniqueness of Christ as the only son of God and the uniqueness of biblical events. Christ's birth, crucifixion, resurrection, and predicted Second Coming are historical events that are cornerstones of Christian faith. If time is cyclic, it does not necessarily mean that specific events are repeated, but for some people it diminishes the significance of these biblical events, because these occur in but one of some possibly enormous number of cycles.

Not all Scholastics were willing to simply appeal to Scripture or, even less likely, to mystical insight to support positions of faith, and since the lengths of these cycles of time were determined by mathematical methods, some turned to mathematics to refute their existence. John Duns Scotus (c. 1265–1308) is credited with being the first to take this approach.[3] Scotus used the now-familiar incommensurability of the side and diagonal of a square and argued as follows: Suppose there are two planets that travel along a square with the same velocity, one moving around the perimeter of the square and the other moving back and forth along the diagonal. (Scotus knew this is not how planets move, but he was trying to show that it is not necessarily the case that planetary cycles are commensurable. Once Scotus proved that it is logically possible for two cycles to not be commensurable, he appealed to the authority of the Bible.) For simplicity, assume that the length of each side of the square is 1 unit, so, by the Pythagorean theorem, the length of the diagonal is $\sqrt{2}$ units. Next suppose that the two square-bound planets are both at a corner, A, of the square at the same time, and that they are both at A at some later time. An analysis of the motions of these two planets is easily carried out using the familiar equation: *distance = rate × time*. When the two planets return to A they will have traveled the same period of time, and as these planets are traveling at the same rate, the above equation implies that they will have traveled the same distance. Thus some multiple of $\sqrt{2}$ units, for example, $m \times \sqrt{2}$, equals some multiple of 1 unit, for example, $n \times 1$. It follows that $\sqrt{2} = n/m$, which contradicts what the Pythagoreans had shown eighteen hundred years earlier. From this Scotus concluded that these two planets will never be in the same position at the same time again, and so time need not necessarily be cyclic.

The shortcoming of Scotus' reasoning, even if we ignore the geometrically improbable shapes of his planets' orbits, is that the two planets are moving with the same velocities. But even if Scotus had been more daring and allowed one planet to move twice as fast or one-third as fast as the other, he still would have been able to conclude that the planets will never return to the same relative positions because the lengths of their orbits are incommensurable.

Scotus chose his example because he knew that if the spheres that determine celestial motions have radii that are commensurable distances, and if they all move at commensurable speeds, then the motions in the sky are cyclic. In other words, Pythagorean perfection in both the structure and physics of the heavens means that time could be cyclic. Thus, despite Scotus' argument using mathematically imperfect planetary paths, other Scholastics sought to disprove the possibility that the sky follows a precisely repeating pattern without appealing to mathematical constructions.

According to Grant, Johannes de Muris (c. 1290–c. 1351) addressed whether or not the velocities of any two planets must be commensurable in his *Quadripartitum numerorum* (Fourfold Division of Numbers [1343]).[4] De Muris correctly argued that if two planets move with the same velocity around circles with incommensurable radii, then the two planets will never return to the same relative positions. Like Scotus before him, de Muris presented no evidence to support his argument; he simply gave the logical argument that it is possible for two planets to have incommensurable cycles and, therefore, for time not to be cyclic.

Even if these arguments did not completely dispel the idea that time could be cyclic, they did make incommensurability more acceptable. The existence of incommensurable geometric magnitudes, or incommensurable natural velocities, in the heavens absolved the Scholastics of having to defend the biblical account of time against the significance of the evidently commensurable cycles in the sky.[5]

ANOTHER VIEW OF CYCLIC TIME: HISTORICAL TIME

Poetry, like the world, may be said to have four ages, but in a different order: the first age of poetry being the age of iron; the second, of gold; the third, of silver; and the fourth, of brass.

— Thomas Love Peacock, "The Four Ages of Poetry" (1820)

Mathematical refutations of repeating cycles in the heavens did not end the speculation that time, or at least some aspects of it, were cyclic. Thomas Love Peacock (1785–1866) found cycles in the evolution of poetry within any particular culture; Oswald Spengler (1880–1936), in the rise and fall of any civilization; and William Butler Yeats (1865–1939), in the course of human existence. Each of these writers offered fairly detailed descriptions of these cycles.

Peacock's cyclic view of poetry was a reversal of the Greek idea that because time is cyclic, and so there is a period of rebirth, there is a natural decline in any culture. A civilization will progressively move from its original mythical golden age through equally mythical silver, bronze, and then iron ages. In Peacock's theory, poetry does not decline but becomes more refined, and eventually pretentious; he offered a description for each period. For example, in the iron age poets are "rude bards [who] celebrate in rough numbers the exploits of ruder chiefs, in days when every man is a warrior, and when the great practical maxim of every form of society [is] to keep what we have and to catch what we can." In English poetry Peacock gave as examples "the rhymes of minstrels and the songs of the troubadours." The golden age of poetry "begins when poetry begins to be retrospective; when something like a more extended system of civil polity is established" as in the poetry of Shakespeare. In the silver age, poetry "is characterized by an exquisite and fastidious selection of words, and a laboured and somewhat monotonous harmony of expression: but its monotony consists in this, that experience having exhausted all the varieties of modulation, the civilized poetry selects the most beautiful." Alas in the last period, the brass age in which Peacock thought he was living, poetry rejects "the polish and the learning of the age of silver, and taking a retrograde stride to the barbarisms and crude traditions of the age of iron, professes to return to nature and revive the age of gold. This is the second childhood of poetry."[6]

Just as Peacock saw pretension in the poetry of his time, particularly in the poetry of William Wordsworth (1770–1850), Spengler and Yeats thought they saw increasing chaos in their time and incorporated that into their theories. They both offered their theories in 1925, Spengler in his book *The Decline of the West*, and Yeats in his book *A Vision*.

Spengler's cyclic understanding of civilization is mythical and imprecise, but it describes how all nations pass through three stages:

1. the emergence from barbarism wherein the society invents gods and heroes
2. a heroic period when human nobility and virtuous behavior is understood to be divine in origin
3. the emergence of reason as a foundation for society and its laws

This last step allows for the eventual decline of the nation because it allows society to become comfortable and to pursue luxury. During this period, the society is either conquered by another or saved by the emergence of a strong leader who steers the culture toward its earlier values. Either way, the cycle begins anew.

Spengler's view of civilization was not so much tied to specific stages as to its *destiny*. Roughly speaking, civilization is the destiny of a culture and decline is the destiny of a civilization. With this model in mind, Spengler concluded his two-volume study with the following paragraph:

> For us, however, whom a Destiny has placed in this Culture and at this moment of its development—the moment when money is celebrating its last victories, and the Caesarism that is to succeed approaches with quiet, firm step—our direction, willed and obligatory at once, is set for us within the narrow limits, and on any other terms life is not worth living. We have not the freedom to reach to this or to that, but the freedom to do the necessary or to do nothing. And a task that historic necessity has set *will* be accomplished with the individual or against him.[7]

Yeats' cyclical view of history is also mythical, but in contrast to Spengler's, the lengths of the cycles are exact. For Yeats, there was a large cycle to history based on the 36,000-year Platonic perfect year; this cycle is subdivided into 2,000-year periods, each of which has a mythical beginning and end. The first 2,000-year period began with the first "annunciation," which Yeats described in his poem "Leda and the Swan" (1928).

This first period had ended with the birth of Christ, and nineteen centuries later the second 2,000-year period was approaching its end. Yeats wrote in "The Second Coming" his oft-quoted lines, "Things fall apart; the center cannot hold; / Mere anarchy is loosed upon the world".[8]

ETERNAL TIME

[The] sum total of the universe is everlasting, having no space outside it into which matter can escape and no matter that can enter and disintegrate it by force of impact.

— *Lucretius*, The Nature of the Universe *(50 B.C.)*

We saw in chapter 3 that Lucretius offered a rebuttal to Aristotle's conception of the universe as being finite, and because Lucretius believed the universe to be infinite he concluded that the universe is everlasting. It follows, of course, that future time must be infinite.

Although Aristotle's theory of infinity led him to conclude that the universe must be finite, implicit in his natural philosophy is the infinitude of time in both directions. Aristotle's conception of motion led him to believe in the infinite divisibility of both time and space. But the infinite divisibility of time presented Aristotle with a philosophical difficulty: What is meant by the present? Because Aristotle believed there are an endless number of moments between any two events, how was one to understand *now*. After all, why are not all of the unlimited number of moments between the past and the future jammed up against one another excluding any particular sense of now? Aristotle's solution to this philosophical dilemma was both simple and brilliant: "Now" is precisely that moment that separates the past from the future. One consequence of this conception of the present is that time must be infinite by addition into the future; a more troublesome consequence is that time must also be infinite into the past.

First, consider the reason there must be an infinite past. Since now is defined to be the moment separating the past from the future, at any given moment there must have been an earlier moment. That is, there could not be a first moment because that moment would have then been the present, separating the past from the future. Just as there was yesterday, there was a day before yesterday and a day before the day before yesterday. This conclusion is contrary to our intuition, in our daily experiences processes have a beginning, perhaps even an initial cause, and if there has been an infinite past it is at least logically possible for there to be a current process that had no beginning.

Aristotle's proof that there must exist an infinite past was not the only one given by either him or his later adherents. The contemporary historian Herbert Davidson has surveyed these proofs and pointed out that many of them are based on either the nature of the world or on the nature of God; he further categorized the proofs based on the world into two types—those intended to establish the eternity of the universe and those intended to establish the eternity of matter (or at least some material that existed before the creation of the universe).[9] Of the proofs seeking to establish that the physical universe must be eternal, six can be traced back to Aristotle. One of these, the one above, relies on the nature of time; of the others, four rely on the Aristotelian principle that matter cannot be created, and one on the nature of motion.

Davidson also isolated three forms of the proof of the eternity of the physical universe based on assumptions about the nature of God; only the first form is discussed here. These proofs argue that no given moment in the undifferentiated pre-Creation universe could have suggested itself to the Creator as the proper moment for creating the universe—and there are two versions of this argument associated with the topics of this book.

One version leads back to Parmenides' metaphysically infinite reality. In this worldview, no moment is distinguishable from any other, so no moment could be the one where Creation begins. Davidson paraphrased the second version, attributed to the fifth-century Neoplatonist philosopher Proclus: "If the world were created, the creator would, up to the moment of creating the world, have been a 'potential creator,' and something would have had to 'activate' him. But the activating factor would, before inducing the creator to create the world, have been a 'potential' activating factor, and hence would have stood in need of a prior factor to activate it as well."[10] This led Proclus to conclude that assuming the Creator created the physical universe leads to an infinite regression of causes—and this is taken as an absurdity.

PROOFS IN SUPPORT OF A FINITE PAST

Aristotle ... may perhaps ask us how we know that the Universe has been created. ... We reply, there is no necessity for this ... but only its possibility. ... When we have established the admissibility of our theory, we shall then show its superiority.
— *Maimonides,* Guide for the Perplexed *(12th century)*

In his *Guide for the Perplexed*, Maimonides discussed God's role in the creation and design of the universe. According to Maimonides, it is easier to explain the intricate motions of the heavenly spheres through an appeal to a creator of the universe rather than through an application of Aristotle's natural philosophy. Yet others attempted to establish that the universe had to be created, not through appeal to the structure or contents of the universe, but by refuting the possibility that time could be infinite into the past, which would imply that the biblical account of Creation could not be correct.

John Philoponus (c. 490–570) offered several arguments against an infinite past; A. W. Moore summarized one of these arguments in his book *The Infinite*: "However many men there were before Socrates, there have been more by now.... So if the numbers here were infinite, one infinity would be greater than another. But this is absurd. So the numbers must be finite. The world must be finitely old."[11] Before we examine this proof, it is important to recognize one of its assumptions: However long the world has existed it has been inhabited by humans. Of course, the biblical account has the world and humans created within the same week. Indeed, it was assumed that the earth existed *for* humans.

Although Philoponus' argument predates the Scholastics by seven centuries, his argument is a quintessential Scholastic one: It invokes one Aristotelian principle, that there cannot be a lesser and a greater infinity, to reject another Aristotelian principle, that time must be infinite into the past. Not all men would ever actually exist at the same moment. There would never be an existing infinitude of men, but there would be an existing infinity of souls. Medieval theologians did not see human souls as immaterial bodies without position, but as eternal remnants of human existence forever residing in a material heaven, hell, or possibly purgatory; so, for the Scholastics, an infinite past, producing more and more dead souls, would present an existing infinitude.

Even if they were willing to accept that God could allow for the existence of such an infinite collection, the existence of infinitely many souls could be dismissed through an appeal to another Aristotelian principle: If time were infinite into the past, there would now be an existing infinitude of human souls (residing in some region of the material universe). A few years from now, after more deaths, there would be more souls and so a larger infinitude of souls; the existence of these souls would violate the fundamental principle that the whole is greater than the part.

The principle that there cannot exist an infinite collection was so strongly held that it even applied to immaterial entities. Bonaventure (1221–74) offered two arguments against an infinite past based on the observation that if time were infinite into the past, the sun would have by now completed infinitely many revolutions around the earth. Following this observation Bonaventure rejected an infinite past by calling upon two Aristotelian principles. The first of these is that an infinite process cannot be completed. Thus, according to Bonaventure, it is impossible for infinitely many revolutions to have been completed; so if there were an infinite past the present would never have been reached. The second Aristotelian principle is the now very familiar one that there cannot be two sizes of infinity. This principle is violated because, if the sun has by now completed infinitely many revolutions of the earth, then, after tomorrow, it will have completed a larger infinitude of revolutions.

SCRIPTURAL TIME

'Tis too late to be ambitious. The great mutations of the world are acted, or time may be too short for our designs.... We whose generations are ordained in this setting part of time, are providentially taken off from such imaginations.
— *Sir Thomas Browne*, Hydriotaphia *(1658)*

Augustine wrote in *The City of God* that time came into existence with God's creation of the cosmos and will come to an end. According to Augustine, time is linear, moving inexorably from Creation to the Last Judgment, and for Augustine this brief period was divided into six stages: Creation to Deluge, Deluge to Abraham, Abraham to David, David to Babylonian Captivity, Babylonian Captivity to the Birth of Christ, and the Birth of Christ to the Last Judgment. Time itself ends after the Last Judgment, because time depends on change and there is not any change in the eternal existence following death.

Not only was it generally recognized that time was not infinite into the past, the earth was thought to be fairly young. Early in the seventeenth century, Archbishop James Ussher (1581–1656) gave the best-known estimate for the date of the beginning of the world—Sunday, October 23, 4004 B.C. To discover this precise date Ussher worked backward through the three cycles, which were the basis for the calendar of his day, the Julian calendar. These cycles were a twenty-eight-yearlong sun-cycle, a nineteen-yearlong lunar cycle, and a fifteen-yearlong indiction cycle. His goal was to find the precise date for some biblical event and then to use the narrative from the Bible to work backward from that point to Creation. Ussher was able to give a date for the end of Nebuchadnezzar's reign and the beginning of Evilmerodach's. He then counted backward, using generations from the Bible, to arrive at 4004 B.C. (Ussher also calculated that Adam and Eve had been expelled from paradise on November 10, 4004 B.C., and that Noah's ark had settled on Mount Ararat on May 5, 1491 B.C.).

In *A History of the Warfare of Science with Theology in Christendom* (1896), Andrew White described a more precise estimate for the date, and time, of Creation. According to White, in 1644, before Ussher published his dates, John Lightfoot, the vice chancellor of the University of Cambridge, claimed that as a result of "his most profound and exhaustive study of the Scriptures," that "heaven and earth, centre and circumference, were created all together, in the same instant, and clouds full of water, [and that] this work took place and man was created by Trinity on October 23, 4004 B.C., at nine o'clock in the morning."[12]

It is not difficult to imagine that there might be an endless future, by simply extrapolating from our daily experience of time, which appears to occur as a succession of moments one after another. Indeed, it is more difficult for us to imagine that time might stop at some point in the future than it is for us to imagine that it continues. Aristotle coupled this observation with the one that says every process or effect must have a cause to conclude time must be infinite by addition into the future; his physics did not allow for a mechanism that could stop time or destroy the universe.

But as the above quotation from Sir Thomas Browne (1605–82) reveals, even in the seventeenth century there was a pervasive view that not only was time not infinite into the future but also that the end of the world was not too far off. No proof was necessary. There was a widely held belief that external evidence supported the notion that the end was near. One source of evidence was the earth's topography, what Marjorie Hope Nicolson has called "mountain gloom."[13] This is the notion that mountains on the surface of the earth are evidence of more than just a lack of perfection; they are evidence of decay. And if there is decay in God's harmonious universe, then we must be near the end of time.

That mountains are deviations from the expected smoothness of the earth's sphere is, of course, a Pythagorean view, but it is one that predated the sixth century B.C. According to etymologists, early Greeks gave names to mountains indicating fear or wildness, although these same Greeks must have also viewed mountains with some reverence since their gods resided on Mount Olympus. The Romans continued this mixed view of mountains, seeing them as being both hostile and protective (the Alps separated Italy from the barbarians to the north).

There is also a theological tradition, probably a consequence of Neo-platonist Pythagoreanism, that mountains are somehow imperfections in need of explanation. This belief takes two forms: One that the earth was originally smooth and round and the other that God created mountains and valleys when he created earth. This latter position is that of Calvin (1509–64), and was later held by Milton and Salluste du Bartas. In this tradition, on the third day of Creation God created earth and, on earth, the Garden of Eden. Eden did not encompass the entire planet but was some small portion of the surface. Calvin even provided a map of Eden complete with mountains. According to Milton:

> God said,
> Be gather'd now, ye waters under heaven,
> Into one place, and let dry land appear.
> Immediately the mountains huge appear
> Emergent, and their broad bare backs upheave
> Into the clouds, their tops ascend the sky.[14]

Salluste du Bartas' description of the emergence of the earth from chaos on the third day of Creation begins with mountains "whose high horned tops" are submerged under the seas:

> Untill th' All-Monarch's bounteous Majesty
> (Willing t' enfeoff man this World's Empery)
> Commanding *Neptune* straight to marshall forth
> His Floods a-part, and to unfold the Earth.[15]

Luther (1483–1546) espoused the other, darker theological view of mountains; the tradition that led Browne to see decay everywhere and to feel the end of time was at hand. The deluge, the flood that Noah rode out on the ark, climaxed man's sin. This flood led to the existence of mountains. A companion theory to Luther's is that God created mountains when he expelled Adam and Eve from Paradise. Both of these views associate the existence of mountains with humanity's failure, that is, with sin. This position manifested itself in the poetry of Andrew Marvell (1621–78) and of Donne. Marvell wrote:

Here learn ye Mountains more unjust,
Which do abrupter greatness thrust,
Which do, with your hook-shoulder'd height,
The earth deform and heaven fright.[16]

In "An Anatomy of the World: The First Anniversary" (1611), the often pessimistic John Donne called mountains "warts . . . in the face / Of th' earth."[17] Just as mountains revealed decay on earth, evidence of decay was also visible in the heavens. A new star seemed to form in 1572, and another in 1604. Then Galileo discovered what could be taken as further evidence for heavenly decay; with his telescope he saw mountains on the moon and spots on the sun.

POSTSCRIPT: TWO PARADOXES OF TIME

We end this chapter with two paradoxes concerning infinite time. The first, according to Moore, is credited to the philosopher Ludwig Wittgenstein (1889–1951) and considers one consequence of time being infinite into the past.[18] We adapt Wittgenstein's original paradox for our purposes. Suppose someone sets about the task of reciting the counting numbers and on the first day begins: "One, two, three, four." They will not be able to complete this project in a day, or a week, or even in their lifetime. So, on their deathbed this person stops their recitation with the dying words, "1 trillion, 7 million twelve." A heartbroken relative then takes up the project, and eventually passes the task on to someone else. As generations follow generations, each making its way through more and more of the counting numbers, the reciters will eventually pass 1 billion million and then 1 trillion billion, but they will never reach a last number because there is not one.

When this idea is reversed, imagining someone reciting the counting numbers in reverse order, the existence of an infinite past yields a paradoxical conclusion. One day a man could walk up to you and say, "Listen carefully, we have been carrying on this project for a very, very long time and I need someone to witness that we have finally finished." The man then continues, "Five, four, three, two, one," and cheers ecstatically. You ask him what that was all about and he replies, "We have just finished reciting the counting numbers backward." No person ever said the largest counting number because there is not one, but each and every number had been spoken at some past time. Whichever counting numbers someone had recited in the 1340s, someone in the 1330s would have recited larger ones. This is extraordinary, but if the world (and our species) has an infinite past, this project has no beginning; each generation inherits the task from the one of its parents. It has simply always been.

The second paradox, which Russell described in 1901, relies on the possibility that time will be infinite into the future.[19] Laurence Sterne (1713–68) exploited this paradox in his eighteenth-century novel *The Lives and Opinions of Tristram Shandy, Gentleman*. Sterne's novel is Tristram's telling of his life story, but Tristram's narrative is not sequential. Tristram keeps getting distracted from his story: He reviews histories and events that occurred before his birth and he often comments on the progress he is making in writing his memoir. He does not even tell these stories in chronological order. In book 4, Tristram tells the reader:

> I will not finish . . . till I have made an observation upon the strange state of affairs between the reader and myself. . . . I am this month one whole year older than I was this time twelve-month; and having got, as you perceive, almost into the middle of my fourth volume— and no farther than to my first day's life—'tis demonstrative that I have three hundred and sixty-four days more life to write just now, than when I first set out; so that, instead of advancing as a common writer, in my work, with what I have been doing at it—on the contrary, I am just thrown so many volumes back.[20]

It took Tristram two years to write the interrupted details of the first two days of his life, leading us to believe he will never complete the narrative of his childhood, let alone of his entire life. If Tristram continues writing his life story at the rate of two days every two years, corresponding to the rate of one day per year, he will only be able to write about the first sixty or seventy or eighty days of his lifetime. The paradox is what happens if Tristram were to live forever. Although Tristram will have related an increasingly insignificant portion of his life—after 365 years he will have told us about only his first year—the assumption that Tristram lives forever allows him to tell us about each and every day of his life. By covering one day every year he will be able to describe his second birthday in his 730th year; he will write about his 730th birthday in his $365 \times 730 = 266{,}450$th year. Although the narrative will never be completed, Tristram will eventually write the details of any particular day of his life.

7 Medieval Theories of Vision and the Discovery of Space

[The] space in which we live and act is not what is treated in art at all. The harmoniously organized space in a picture is not experiential space, known by sight and touch, by free motion. . . . It is an entirely visual affair. . . . This purely visual space is an illusion.
— *Susanne Langer,* Feeling and Form: A Theory of Art *(1952)*

Einstein's twentieth-century physical theories connect four of the most enigmatic of all concepts: space, time, gravity, and light. Although each of these abstractions is essential to our understanding of the natural world, and we cannot even imagine a world without these components, each one had to be discovered. Before Newton gave his famous law for the gravitational attraction between two bodies, objects simply sought their natural position, Aristotle's "place" for the object. Before the development of reliable timepieces, time was not continuous, and it did not flow from the past to the future with the present wedged in between; time was a succession of events. Before the Greeks asked how it is that we see things, light was an attribute of fire, and not the mechanism for vision. And attempts to explain vision with light assisted in the artistic discovery of space.

Early theories of vision did not involve light, but some sort of visual force. To the *atomists*, all objects in the physical world emit *eidola*, picture-like effluxes that enter the eye. Each object carries with it its images; our eyes are passive receptors. The early Pythagoreans criticized this process, called *intromission*, asking, how could the image of a large object enter our small eyes? The Pythagoreans offered an alternative theory of vision—they believed that our eyes emit a visual force that produces the images we see. They did not offer an explanation for how the visual force striking an object produced an image at its source, but Plato addressed this issue. Plato claimed that the fire coming from our eyes interacted with a force being given off by the object; this interaction transmitted an image of the object back into the eye. It was Aristotle, the empiricist, who brought light into the theory of vision. Aristotle maintained that the presence of light caused the medium between the viewer and the object, either air or water, to produce an image of the object. This is very close to the modern theory that our eyes collect light rays bouncing off the object (and our brain processes this information to form an image).

This chapter examines the connection between the geometric underpinnings of how vision is imagined to work and how artists paint a scene, design a building, or more generally represent space. The goal of an artist, at least of a nonabstract one, is to represent a scene or object so that a viewer will understand what is being represented. To accomplish this, the artist must take into account not only what is known about how we see things but also how the artist thinks the viewer will respond to his or her representation. And depending on which theory of vision the artist has in mind, the representation may or may not look convincing to the modern eye.

EUCLID'S OPTICS

The theory of vision given by Euclid in his *Optics* dominated intellectual thought throughout the Middle Ages. Euclid's geometric treatment of vision, with some minor modifications, was widely accepted in Europe until the Renaissance. Basic to Euclid's theory was his introduction of the visual cone, a cone formed by *visual rays* emitting outward from a fixed point inside the eye. Euclid's theory of how these visual rays yield mental images of the objects they encounter is not important to understanding how Greco-Roman and then medieval optics influenced artistic representations of space. The most artistically influential aspect of Euclid's theory is its geometric explanation of how we perceive the size of an object within the visual cone.

In Euclid's theory, the collection of visual rays falling on an object, for simplicity of explanation the vertical line segment A (below) forms an angle a at their source:

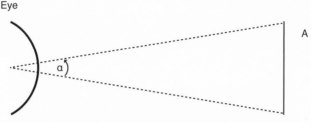

FIGURE 7.1. In Euclid's theory our perception of the size of an object is determined by its visual angle.

This is the *visual angle* associated with the object. Euclid's basic principle is then:

Two objects that give rise to the same visual angle
are perceived to have the same size.

Euclid's visual-angle equivalence for the apparent sizes of objects agrees with our commonsense notion that the farther away an object, the smaller it appears. This connection between distance and apparent size follows from Euclid's geometry and an application of his basic principle. Consider Figure 7.2:

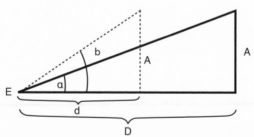

FIGURE 7.2. The visual angle of an object is reduced if the object is moved farther from the eye.

In this drawing the viewer's eye is represented by the point E, and the object being observed is the line segment A, which is first perceived at a distance d from the viewer, then at a larger distance D from the viewer. When the segment A is located at the shorter distance d from the viewer's eye, it has an associated visual angle of b; when the segment A is moved farther from the viewer, to a distance D, then the new visual angle of A, the angle a, must be smaller than b. It follows from

Euclid's basic equivalence that once A has been moved, it will appear to the viewer to be smaller.

Although Euclid's connection between the visual angle of an object and its apparent size is aligned with common sense, it is contrary to a fundamental principle of perspective painting. Consequently, the acceptance of Euclid's optics possibly forestalled the development of a method for convincingly representing three-dimensional space for fifteen centuries. The precise difficulty with Euclid's emphasis on the visual angle lies not in how the apparent size of an object changes when it moves away from the viewer horizontally, but when it moves in a fixed plane perpendicular to the central ray in the visual cone. If object A is first viewed directly opposite the viewer, and is then displaced upward, A would then be farther away from the viewer, and hence have a smaller apparent size (the same reasoning works if A is shifted laterally). Before examining how this simple observation violates one of the fundamental principles of perspective painting let's consider the impact of Euclid's optics on the artists and architects who may have tried to employ its principles.

EUCLID'S OPTICS AND ARTISTIC REPRESENTATIONS

An artist trying to represent a scene, or object, while thinking of size in Euclid's terms, must make a choice, because the visual angle and apparent size equivalence posits a discrepancy between what is thought to be seen and what measurement shows exists. A simple example illustrates this conflict. According to Euclid's theory, if you were to stand at a fixed distance from a row of fence posts, the posts that are shifted to the left or right will appear to be smaller than the middle pickets. Thus, the row of fence posts will appear to you as

FIGURE 7.3. An exaggerated view of how an ordinary fence should appear to an observer in Euclid's theory.

Although measurement would show that all of the posts are the same height, the artist would imagine that the viewer would see the fence as we have drawn it above. The artist must decide whether to represent the fence posts as they appear or to draw a version of the fence posts so that what is thought to be seen will be interpreted correctly. The latter choice, of representing an object so that it will be correctly interpreted, explains two different aspects of Greco-Roman and medieval art—*reverse perspective*, wherein a rectangular surface is drawn with a wider back edge, and *entasis* in columns, the construction of columns with a slight bulge in their center. Each of these two stylistic innovations results from a different method for representing an object so that the viewer will correctly interpret it.

First, suppose a carpenter, who accepts Euclid's basic principle, is hired to construct a fence and his employer insists that the fence look perfect. The carpenter interprets his employer's instruction to mean that he wants all of the fence posts to appear to be of the same length. Since the farther a fence post is from the center post, the smaller it will appear to a viewer, the craftsman constructs the fence with fence posts whose lengths increase with their distance from the center post:

FIGURE 7.4. How an artist might represent an ordinary fence in order for the viewer to understand that the fence is properly constructed.

Then, according to Euclid's theory, the fence will appear to have fence posts with equal lengths.

An artist representing any object could import the reasoning of the carpenter into a painting. The carpenter's reasoning as represented in Figure 7.4 would lead the artist to employ reverse perspective, as, for example, in representing the presumably rectangular top of the well in this detail from a Byzantine icon.

Alternatively, if the carpenter were to construct the fence taking into account the expectations of the viewer that the fence is being con-

PLATE 7.1. *The Woman and the Well* (detail from *Jesus and the Samaritan Woman at the Well*). Byzantine icon, late 16th century. Overall size 38.5 × 48 cm. Paul Canellopoulos Collection, Athens, Greece. Photo: Erich Lessing / Art Resource, New York.

structed so as to be correctly interpreted, it would have a different shape than the fence in Figure 7.4. Such a viewer would expect that since the end fence posts should appear shorter than the center fence post, they must really be taller than the center post; the carpenter would have to compensate for both the effects of distance on size, and for the adjustments the viewer will assume have been made. This means that the end fence posts must be made shorter than the center post—the viewer will know that the fence posts are all of equal lengths.

This is not an entirely theoretical discussion; these principles have been applied in both art and architecture. For example, if a tall, rectangular door is constructed without any adjustments the viewer will assume the door has been constructed so that it appears rectangular. The viewer will realize, or expect, that since the top of the door should appear to be narrower, as it takes up a smaller visual angle than the base, the door must really spread outward toward the top. So, an architect, or artist, must compensate for both the effects of distance on size and for the adjustments the viewer will assume have been made. This logic leads to the representation/construction of a tall, relatively narrow door with a top that is narrower than its bottom. The viewer will imagine that this door is rectangular; its top edge is smaller because it is farther away.

The first-century B.C. Roman architect Vitruvius (c. 70–c. 25 B.C.) prescribed how to adjust the measurements for a doorway in a temple in order that it appear to be rectangular. After describing how to determine the width of the base of a doorway from its height—if the height of a doorway is divided into 12 parts the base should be 5½ of these parts wide—Vitruvius offered the precise scaling to be used (his assumption being that the farther an object is from the viewer the smaller the necessary adjustment): "At the top, [the door's] width should be diminished, if the aperture is sixteen feet in height, by one-third the width of the door-jamb; if the aperture is from sixteen to twenty-five feet, let the upper part of it be diminished by one-quarter of the jamb; if from twenty-five to thirty feet, let the top be diminished by one-eighth of the jamb. Other and higher apertures should, as it seems, have their sides perpendicular."[1]

Examples of Vitruvius' scaling factors are given below:

HEIGHT OF DOOR (in ft)	RATIO OF DOOR'S TOP WIDTH TO DOOR'S BOTTOM WIDTH
< 16	2:3
16–25	3:4
25–30	7:8
over 30	1:1

This formula tells an architect who wants to design a twenty-four-foot-tall door, which must necessarily be eleven feet wide at its base according to Vitruvius' earlier rule, that the two sides of the door should not be perpendicular to the floor but slant inward forming an eight-and-a-quarter-foot-wide opening at the top of the door.

The contemporary art historian Kim Veltman relates the details of an art competition, the goal of which was to design the most appealing statue of Minerva.[2] The statue was to be placed on a high pillar so that it would be viewed from below, with viewers looking up at both the pillar and Minerva. According to the story, two sculptors, Phidias and

Alcamenes, entered the competition. When viewed at ground level, Alcamenes' statue was seen to be a beautiful and perfectly proportioned figure; Phidias' Minerva was distorted, her head and shoulders were too small for the rest of her figure, as would be the top edge of a door constructed according to Vitruvius' guidelines. After the two statues were seen on the pillar, Phidias' small-headed Minerva was the more appealing; he won the contest.

The idea that an artist or architect must take into account the viewer's expectation that some adjustment has been made, in order that an object be interpreted correctly, can also lead to the principle of entasis in columns, wherein a tall, rectangular column is designed with a slight outward bulge in its middle. To follow this reasoning we begin with the assumption that we are viewing a column whose middle portion is roughly at eye level. Suppose the column is constructed in the form of a simple, tall rectangle. Since both the top and base of the column are farther away from the viewer than the center, the viewer will assume that the column has been constructed to appear as a rectangle and thus must be wider at both the top and bottom than in the middle. To adjust for this, the column should be constructed to be narrower at both the top and bottom, that is with a bulge in the middle, in order that the viewer understand that the column is a rectangle.

There is another, less mathematical and psychological, explanation as to why Greco-Roman columns were constructed to be biconvex: when designers looked at older buildings, which had originally been constructed with rectilinear columns, the weight of the structure had, over time, compressed the columns to produce a bulge in each. These later architects simply copied this style.

ALHAZEN'S THEORY OF VISION

In the eleventh century, Alhazen (965–1039) offered a theory of vision that at first seems similar to Euclid's but is significantly different in ways that allowed for the development of single-point perspective painting. The conceptual shift concerns the medium of vision itself. Alhazen assumed that the visual rays emanate from the objects in the physical world, instead of from within the eye. His reasoning is straightforward: When we look at a very bright object such as the sun we experience pain. This seems to be inconsistent with the model of visual rays emanating from the eye, unless one wants to attribute to all physical objects some sort of reflectivity principle, such as that bright objects simply have more of this mysterious property than dim objects. But some objects, a flickering candle for example, vary in brightness from moment to moment. This means that this mysterious property would have to be variable.

Alhazen's understanding of the mechanism of vision does have one difficulty that he had to overcome—if all objects are emitting visual rays in all directions, then our eyes should be overwhelmed by the barrage of images. Alhazen's answer to this objection, which contains in it one of the two ideas that led to the discovery of the single-point perspective method of painting, is that the lens of the eye only admits those rays that meet it at a right angle. By the eleventh century, the basic physical structure of the eye was understood, and it had been known for a millennium that refracted light had a diminished intensity. Alhazen took these elements into account in developing his theory of vision, which involves the following idealized view of how light enters the eye.

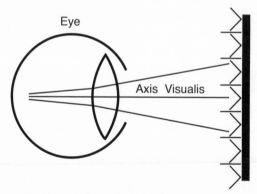

FIGURE 7.5. In Alhazen's theory the light along the *axis visualis* is not refracted, so vision is clearest along that axis.

The only light that is not refracted, and so not diminished in intensity, is the light reaching the eye along the *axis visualis*. Thus, there is a clearest line of sight, and vision becomes less clear the farther the visual rays are from that axis.

The existence of the *axis visualis* is but one aspect of Alhazen's theory of vision relevant to the impressive fifteenth-century innovations in representing three dimensions on a two-dimensional surface. The other, which is not entirely unrelated to the first, is Alhazen's distinction between immediate and contemplative perception. Immediate perception is a person's initial, relatively vague impression of an object or scene. Contemplative perception is clearer; through a process Alhazen labeled *certification* a more detailed and accurate mental image of a scene is created over time. It was already understood that if light, or visual rays, did not travel instantaneously, their velocity was imperceptibly fast. So Alhazen imagined the time required for certification to be relatively short, but not immediate.

Thus, Alhazen's explanation for how someone observing a canvas creates a mental image of the painting appeals to his conclusion that sight along the *axis visualis* provides the clearest, and therefore most powerful, image. To obtain the clearest possible image, the observer should remain stationary, so that the image is formed through contemplative perception. Giotto di Bondone (c. 1267–1337) seems to have been the first artist to understand that painting a scene as if the viewer were to be in a fixed position would allow for a more believable representation of three-dimensionality. One only needs to compare his *Meeting at the Golden Gate* with other paintings of the period to appreciate Giotto's almost geometric representation of space and depth.

PLATE 7.2. *Meeting at the Golden Gate*, 1304–6. Giotto di Bondone (c. 1267–1337). Scrovegni Chapel, Padua, Italy. Photo: Scala / Art Resource, New York.

Although Giotto was the first painter to sense the importance of painting a scene as if the viewer remained stationary, the first artist to understand the geometry behind the depth artists began to achieve in the fifteenth century was Filippo Brunelleschi (1377–1446). In the early 1420s Brunelleschi is said to have demonstrated the power of assuming the painting is being viewed from a preferred spot in establishing an impressive three-dimensionality in a two-dimensional painting. Brunelleschi's demonstration occurred on the Piazza del Duomo, in Florence, and involved a perspective painting with a small hole in the canvas at the vanishing point. A mirror was held parallel to, and facing, the painting. People on the plaza were invited to look through the back of the painting and observe its reflection in the mirror. Why? Because Brunelleschi's setup effectively forced the observers to view the painting from a precisely determined fixed point, which allowed for the greatest appreciation of the painting's geometric framework. This framework was formalized a decade after Brunelleschi performed his experiment by Leon Battista Alberti (1404–72) in his influential book *De Pictura* (1435).

In *De Pictura*, Alberti settled many of the geometric technicalities that must be incorporated into a painting to provide a convincing representation of three-dimensional space, and all of his geometric conclusions follow from his unifying conception for what a painting is. As the contemporary art historian Lew Andrews so succinctly puts it in his book *Story and Space in Renaissance Art*, Alberti "likens a painted picture to an open window: A picture, in his view, should be made to seem as if it were a pane of transparent glass through which we look into an imaginary space extending into depth."[3] Another way to state Alberti's idea is to imagine that the painting is a fixed pane of glass, which is called the *picture plane*, interposed between the viewer and the scene. The two central geometric tenets of the single-point perspective method follow from this conception of a painting as a picture plane:

1. objects of the same size in a plane parallel to the picture plane will have the same size in the painting, and
2. any two parallel lines that are perpendicular to the picture plane will appear to converge toward each other.

This first principle is contrary to Euclid's equivalence of the apparent size of an object and its visual angle. The second principle implies that the artist must choose a *vanishing point* for the painting and all parallel lines drawn toward the horizon must converge to this point. Both of these principles follow from an application of Euclidean geometry; the imposition of the picture plane between the viewer and the scene is the new idea that presents the artist with the necessary theoretical tools. But to employ this idea in its purest form, an artist must become a bit of a geometer. In order to give, for example, a convincing portrayal of a large, tiled plaza with people positioned throughout, the artist needs to address two fairly subtle geometric questions:

1. How large should the figures appear when placed in intermediate positions between the plane of the painting and the vanishing point?
2. How much should the tiles in the floor be foreshortened to yield an image of a floor of rectangular tiles reaching toward the horizon?

While the solution to the second question is geometrically interesting, it is not germane to the discovery of space. But the solution to the first question is relevant because it clearly illustrates the difficulty with Euclid's visual angle theory. It is easier to answer this question if it is reformulated as: How should we represent objects that are measurably the same size, all of which lie in a fixed plane perpendicular to the line of sight of the viewer? The answer to this question depends on three pieces of information: the distance of the object from the picture plane, the size of the object, and the imagined distance of the viewer from the picture plane.

To see why the painted size of an object depends on each of these numbers, suppose the viewer is standing ten feet from the imagined picture plane and he is observing an object, w, that is twenty feet beyond the picture plane. (In Figure 7.6 x denotes the apparent size of the object w.)

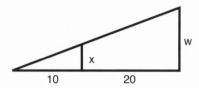

FIGURE 7.6. How large an artist should draw an object depends both
on the distance of the presumed viewer from the canvas and the
imagined distance of the object from the picture plane.

It is easy to relate the numbers by using the observation that the two triangles have the same shapes, because their angles are all the same, thus the ratios of corresponding sides are equal. This yields

$$x/w = {}^{10}\!/_{(10+20)} = {}^{10}\!/_{30} = 1/3.$$

If we multiply through by w we find that $x = 1/3\ w$; so w appears to be one-third as large as it is. If we double the distance of w from the picture plane, so its distance from that plane is now 40 feet, we again use similar triangles to find

$$x/w = {}^{10}\!/_{(10+40)} = {}^{10}\!/_{50} = 1/5.\text{ Thus } x = 1/5\ w$$

so w appears to be one-fifth as large as it is (and should be drawn at one-fifth of its true size).

From this small calculation, we see that if we imagine doubling the distance of an object from the picture plane, its size is not halved (in this example it decreases by 40 percent). If we double the distance of w from the viewer, while leaving the picture plane fixed, then the size to the object is halved. Thus the two distances, the distance of the viewer from the picture plane and the distance of the object from the picture plane, must be taken into account to calculate the relative sizes of objects in a painting.

Leonardo da Vinci (1452–1519) devoted a great deal of time to this matter, producing tables for the relative sizes of objects at varying distances from the picture plane. The rule he developed, which is to be applied to compute how large an object should be drawn in order that it appear to be of the correct size, is known as the *inverse-distance law*. Its fundamental principle is easily understood: The size of an object in the picture plane will be determined by its distance from the picture plane measured in terms of the distance of the viewer from the picture plane.

If the object is the same distance from the picture plane as the observer, the object in the picture plane will be one-half of its measured size. If the object is twice as far from the picture plane as the observer, its size will be one-third of its measured size; if the object is three times as far from the picture plane as the viewer its size will be one-fourth of its measured size. We saw in the example above that when the object is four times as far from the picture plane as the viewer then its size will be one-fifth of its measured size.

CONTINUOUS NARRATIVE

Art is either plagiarism or revolution.
— *Paul Gauguin (c. 1880)*

The standard view of Renaissance art, as espoused, for example, by the early twentieth-century art theorist Dagobert Frey is that once the single-point perspective method took hold, the sequential imagery of the continuous narrative only remained as something of a bad habit. Andrews explained that in Frey's theory, having several temporally disjoint scenes within the same frame was "incompatible with the spatial innovations and representational logic of the quattrocento."[4] Frey's argument continues: Once this incompatibility was recognized, artists dropped the technique of continuous narrative. The reason for this is the assumed superiority of the unified, almost photographic space that single-point perspective provided. It is not only the space of the painting that is photographic, but its time as well. A well-executed single-point perspective painting presents a frozen-in-time, geometrically convincing, three-dimensional image. According to Frey, "Simultaneous unity of content in painting is scientifically attained in perspective." Frey then goes on to say, "Simultaneity in perceiving a picture also requires a synchronization of what is represented; by grasping the picture spatially as a unit we also assume the depicted events to be simultaneous."[5]

Assuming Frey is correct, what is to be made of paintings that were produced after the principles of single-point perspective were understood, such as the panel from Lorenzo Ghiberti's design for the east doors of the baptistery in Florence (Plate 7.3) and a painting by Benozzo Gozzoli (Plate 7.4).

PLATE 7.3. *Story of Jacob and Esau* (panel from Gates of Paradise, 1425–52). Lorenzo Ghiberti (1370–1455). Museo dell'Opera del Duomo, Florence, Italy. Photo: Scala / Art Resource, New York.

PLATE 7.4. *Arrival of St. Augustine in Milan*, 1464–65. Benozzo Gozzoli (1420–97). Sant'Agostino, San Gimignano, Italy. Photo: Scala / Art Resource, New York.

Both of these, with representations of scenes taking place at different times, seem to violate the instantaneous, pictorial nature of single-point perspective paintings. The standard theory tells us that these works are simply a holdover from the earlier technique. Andrews has offered another explanation. Andrews' position is that artists never completely accepted that the temporal aspects of a painting are intricately linked to its portrayal of space. This is easily seen in Gozzoli's painting: Augustine is shown in the foreground with a servant helping him remove his riding clothes, kneeling before an Islamic scholar in the background, and being greeted by Ambrose (right).

These portrayals of several events within the full three-dimensional space the single-point perspective style offers us are a break from medieval continuous narrative paintings, where several events are presented on the same panel but not in the same unified space. But this seems to be a strange juxtaposition of representations. The space of the painting must be viewed from a predetermined point and thus be seen as static and unchanging, while the time of the painting is plastic and changing. We are not the first to have noticed this disparity; early in the sixteenth century Leonardo da Vinci investigated the assumed rigidity of the space in a painting. Leonardo and Piero della Francesca (c. 1412–92) before him investigated this from slightly different points of view. They were both interested in the geometric underpinnings of single-point perspective, and specifically in the relationship between the perceived depth within the painting and the believability of its portrayal of space. They both reduced this relationship to the simple matter of the location of the vanishing point in the canvas.

To understand Leonardo's and Piero's discoveries we must think of the vanishing point not as a point on the canvas but as a point located behind the canvas. To explain this idea we return to Alberti's conception of a painting as a pane of glass interposed between the viewer and the scene. If we imagine examining the picture plane between the viewer and the scene from above, then we have the following schematic view of the situation:

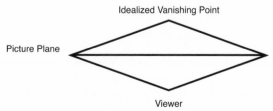

FIGURE 7.7. The preferred position for the viewer of a single-point perspective painting is directly opposite the vanishing point.

In his experiment in the Piazza del Duomo, Brunelleschi exploited the following observation: The preferred location of the viewer of the painting is at the point obtained by reflecting the vanishing point through the picture plane (see Figure 7.7).

If the vanishing point is low on the canvas, the lines in the painting will appear to converge at a point that is relatively nearby and the ideal position of the viewer is very close to the canvas. If the painting is sufficiently wide, the objects in the foreground corners will appear slightly distorted. Moreover, and this is more relevant to our discussion of continuous narrative, our view of the space within the painting is very sensitive to our position. If the viewer moves slightly from the preferred spot close to the canvas, the three-dimensionality of the space will be less convincing.

However, if the vanishing point is very high on the canvas, the lines in the painting will appear to converge at a very distant vanishing point and the viewer's optimal position is relatively far away from the painting. With such a "distant" vanishing point, there will be no evident distortion of objects in the foreground corners. What this means is that if the viewer moves a little, there will be not be a significant change in the scene.

Piero was mostly interested in eliminating any distortion from a single-point perspective painting. He concluded (theorem XXX) that the vanishing point must be placed high enough on the canvas so that its distance from the bottom of the canvas must be at least half the width of the canvas.[6] He also prescribed that the angle at the vanishing point, in the overhead view above, must be less than 90 degrees.

Leonardo was more interested in the viewer's understanding of the painting. For Leonardo, the location of the vanishing point prescribed in Piero's theorem XXX liberates the viewer. While there is a preferred location from which a single-point perspective painting should be observed, that location is not a single point. Due to the distance of the viewer from the canvas, there is some latitude in the preferred location; the preferred viewing spot is within a region, not at a point. Thus the space is not static, something to be passively viewed from only one position. Instead, it is dynamic, we can move through it by changing our position slightly. If the relationship of the viewer to the space of a painting is dynamic, there is no reason that the time of a painting must be static.

Whatever rigidity of method was called for in the early Renaissance, very few artists strictly adhered to all aspects of the prescribed geometry (two who did were Masaccio in his painting *Trinity,* and Piero in his painting *Flagellation*). But one aim of art is to be revolutionary, not to copy from others, and as soon as the principles of single-point perspective were understood, they were challenged. Indeed, rigid, mathematical rules for an artist to follow were not prescribed again until the early twentieth century (see chapter 9).

FRACTURING SPACE AND MULTIPLE POINTS OF VIEWS

The transition from the strict single-point perspective method to modern, abstract art, once it started, was rapid. Although the rules of the single-point perspective painting were immediately violated, these violations did not challenge the single-point perspective's conception of space. Then, in the nineteenth century the very nature of space was questioned, and once this examination began artists were liberated from any Euclidean restraints. The evolution of abstract art took less than a century. The impact of the discovery of non-Euclidean geometries and then the fourth dimension on literature and art are illustrated in the next chapter; for now we restrict our attention to two early examples of artistic affronts to the standard method of representing space.

The first example, below, is from the French painter Edouard Manet (1832–83). (It is considered to be his last major piece.)

PLATE 7.5. *A Bar at the Folies-Bergère,* 1882. Edouard Manet (1832–83). Courtauld Institute Galleries, London. Photo: Foto Marburg / Art Resource, New York.

On first viewing, this painting appears to be a straightforward depiction of a young woman, behind the bar, looking directly at you. Behind the girl is a large mirror, parallel to the bar; in the mirror the viewer sees a reflection of the bar scene as well as the back of the girl. A schematic top view of this scene is

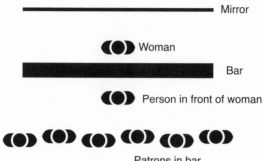

FIGURE 7.8. A top view of the bar in Manet's *Bar at the Folies-Bergère* that does not take into account the image in the mirror behind the woman.

However, assuming that Manet painted this scene according to the commonly accepted principles, the location of the woman's reflection reveals that this cannot be the top view of the scene. The woman's reflection should be directly behind her and not be visible to the viewer; Manet has represented the scene as if the mirror were not parallel to the bar—it is skewed so the woman's reflection appears on the right-hand side of the painting:

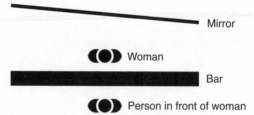

FIGURE 7.9. A top view of the bar in Manet's *Bar at the Folies-Bergère* that takes into account the image in the mirror behind the woman and the apparent relationship between the viewer and the woman.

However, in the painting the bar and the mirror appear to be, more or less, parallel, so to see the woman's reflection where Manet has placed it, the viewer must be standing to the right of the woman:

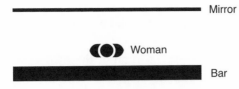

FIGURE 7.10. A top view of the bar in Manet's *Bar at the Folies-Bergère* that takes into account both the image in the mirror behind the woman and the apparent relationship between the bar and mirror.

This placement of the figures is contrary to the view we have of the woman's face and body.

The only solution to this dilemma is to accept that Manet painted *A Bar at the Folies Bergère* as if it is being simultaneously viewed from two different positions, one directly in front of the woman and the other from the right-hand side of the painting.

A more extreme, but less obvious, example of a scene having been painted as if it were being simultaneously viewed from more than one position is Paul Cézanne's *Still Life with Fruit Basket*, below. The objects

PLATE 7.6. *Still Life with Fruit Basket*, 1888–90. Paul Cézanne (1839–1906). Oil on canvas, 65 × 81 cm (R.F. 2819). Musee d'Orsay, Paris. Photo: Erich Lessing / Art Resource, New York.

on the table are slightly distorted, but many of these distortions are resolved if the position of the viewer is allowed to change, depending on which object is being viewed. For example, if the body of the basket were being viewed from the position of the reader, then its handle would be viewed from someplace to the right of the canvas, and although this is harder to see, the large jar next to the basket would be viewed from somewhere to the left of the canvas.[7]

Manet and Cézanne were modern painters, working three centuries after Leonardo's investigations liberated the viewer from having to remain in a fixed place in order to appreciate the three-dimensionality of a painting. The jump from allowing the viewer to move slightly from one position to another to allowing the viewer to be in several positions simultaneously is a huge one. What allowed for this conceptual leap was a two-century reexamination of our assumptions about space, which included the discovery of mathematical geometries different from the one presented in Euclid's *Elements* and Einstein's discovery of the fundamental connection between space and time. These are discussed in the next chapter.

8

The Shape of Space and the Fourth Dimension

You must note this: If God exists and if He really did create the world, then, as we all know, He created it according to the geometry of Euclid and the human mind with the conception of only three dimensions in space. Yet there have been and still are geometricians and philosophers, and even some of the most distinguished, who doubt whether the whole universe, or to speak more widely, the whole of being, was only created in Euclid's geometry; they even dare to dream that two parallel lines, which according to Euclid can never meet on earth, may meet somewhere in infinity.
— **Dostoyevsky, The Brothers Karamazov (1879)**

Early in the eighteenth century, more than two millennia after Eratosthenes measured the length of a single shadow and then determined the shape and size of the earth, the mathematician Carl F. Gauss (1777–1855) is said to have performed another geometric experiment. Gauss placed some of the most accurate surveying equipment of his day on three mountaintops and measured the angles in the triangle formed by the peaks. Gauss, of course, knew the result from geometry that Plato had used in his theory of matter, and that played a hidden role in Eratosthenes' experiment: *The angles in any triangle sum to 180 degrees.* When Gauss found that the angles in his surveyed triangle added up to slightly more than 180 degrees, he could have used the known geometric result to estimate the precision of the equipment. But this had already been determined by measuring angles whose measurements were known. Gauss was not checking the accuracy of mechanical devices; he was testing the hypothesis that Euclidean geometry is a science and that its results are not only mathematically correct but are empirically verifiable. Unfortunately Gauss's experiment was inconclusive because the discrepancy between his result, of just less than

$180\frac{1}{4}$ degrees, and 180 degrees, fell within the known margin of error of his equipment.

ARE EUCLID'S POSTULATES TRUE?

In 1968 the artist Walter de Maria (b. 1935) drew two parallel chalk lines in the flat Mojave Desert. The two lines in de Maria's *Mile-Long Drawing* were twelve feet apart and, contrary to the title of the installation, two miles long. All that remains are photographs of the drawing; in these, the two chalk lines appear to be parallel and yet to converge as they move toward the horizon, meeting at some vanishing point beyond the distant hills. The previous chapter examined how Renaissance painters imported this visual illusion onto canvas to obtain convincing portrayals of three-dimensional space. But despite appearances, parallel lines are not supposed to intersect, and although it is not immediately evident, Gauss's experiment was designed to test this supposition.

The connection between the angles in a triangle and the behavior of parallel lines is fairly subtle, and the total angle measurement of the angles in a triangle is intimately connected with the existence, or nonexistence, of parallel lines. Euclid, of course, never assumed that the angles in a triangle sum to 180 degrees, he deduced this result from his postulates. Furthermore, if the parallel postulate is replaced by the assumption that the sum of the angles in a triangle always equals 180 degrees, then the statement of the parallel postulate can be deduced as a theorem. In this sense the parallel postulate and the sum-of-the-angles result are equivalent; the truth of either one implies the truth of the other.

Between the third century B.C. and the seventeenth century, a fundamental shift occurred in the way geometric objects, and in particular parallel lines, were conceived. This shift can be seen in the poet Marvell's appeal to the nature of parallel lines to express a necessarily unrequited love:

> My love is of a birth as rare
> As 'tis for object strange and high;
> It was begotten by despair,
> Upon impossibility.
>
> . . . .

Unless the giddy heaven fall,
And earth some new convulsion tear,
And, us to join, the world should all,
Be cramped into a planisphere.

As lines, so loves oblique may well
Themselves in every angle greet:
But ours, so truly parallel,
Though infinite, can never meet.[1]

According to Marvell, parallel lines are nonintersecting, existing infinitudes, and by the time of Gauss's experiment this conception of parallel lines was the accepted one.

We cannot test whether two parallel lines are two finite entities that, no matter how far they are extended, will never intersect or whether they are two existing infinitudes. But there are other aspects of Euclid's *Elements* that can be examined, such as whether the angles in any triangle sum to 180 degrees. But instead of climbing mountain peaks, as Gauss did, it is possible to test the truth of this result through a simple experiment involving triangles.

For a triangle drawn on a piece of paper, within the error of your measurement, the triangle's three angles will sum to 180 degrees. Imagine drawing a larger triangle, and instead of drawing it on paper, you draw it on the surface of the earth, perhaps in a large field. If the angles in this large triangle were to be measured, the sum of the angles still would be (as far as could be discerned) 180 degrees. But in an even larger triangle, one that covers most of France, for example, the angles will add to slightly more than 180 degrees. Finally, imagine drawing a huge triangle with one vertex on one of the Galápagos Islands, another vertex on the north shore of Lake Victoria in Africa, and its third at the North Pole. Both the Galápagos Islands and the north shore of Lake Victoria are pretty much on the equator, so the base of this triangle lies, more or less, along the equator. If you were to stand at one of this triangle's equatorial vertices and look toward the vertex at the North Pole your line of sight would be perpendicular to the equator. (This angle will not be exactly 90 degrees because the earth is not a perfect sphere and the triangle's vertices are not precisely on the equator, but it will be close.) This means that this extraordinarily large triangle has a 90-

degree angle at each equatorial vertex. And because Lake Victoria is 30 degrees east of Greenwich and the Galápagos Islands are 90 degrees west of Greenwich, the angle at the North Pole measures 120 degrees; so the sum of the angles in the triangle is 90 + 90 + 120 = 300 degrees, much greater than the expected 180 degrees. (By taking the two equatorial positions farther apart, the angle at the North Pole can be made to be any number less than 180 degrees; so a triangle could have its angles sum to any number less than 90 + 90 + 180 = 360 degrees. Indeed, it is a theorem from spherical geometry that the angles in any triangle drawn on a sphere will exceed 180 degrees.)

The discovery that the sum of the angles of a triangle drawn on the surface of the earth always exceeds 180 degrees does not imply that Euclidean geometry is wrong (or that there is some fundamental flaw in our deductive methods). The assumption that because the earth is curved, figures drawn on its surface need not adhere to Euclidean geometry is a valid one. But the complication to this claim is that experience tells us that the earth is flat, and any attempt to give a simple definition of flatness encounters the same difficulty Euclid did when he wrote:

> A plane surface is a surface that lies
> evenly with the straight lines on itself.

According to this definition, a flat surface is one on which it is possible to draw straight lines. But Euclid's definition of a straight line does not offer any insight into the nature of a flat plane because all it says is that a straight line does not waver, that is, each point on a line lines up evenly with the other points on the line. So the definition of a plane is given in terms of a straight line, and the definition of a straight line depends on our intuition or experience. Although we think we understand what Euclid means by a straight line, if someone were to present us with a curve on a piece of paper we would not have any way of checking whether it is straight or not. (The way we would like to check the straightness of the curve would be to hold a ruler along it, but this leads to an unending digression because we have no way of verifying that the ruler is straight. This dilemma is similar to the one Kant expressed concerning how we can understand the size of any magnitude [see chapter 3].)

The most accomplished of all Greek mathematicians, Archimedes, offered an alternative definition of a straight line that, in principle, can be verified:

A line is the shortest distance between two points.

Using Archimedes' definition of a line, it is possible to avoid Euclid's reliance on *straight* to define *flat*, and on *flat* to define *straight*, however it does rely on measurement. Nonetheless, the real advantage of this shift in point of view is that the concept of a line does not depend on straightness, which cannot be checked, but on distance, which can be checked. So it is reasonable to define the line segment defined by two points on a sphere as the arc on the sphere connecting the two points, which has the shortest length. The line defined by two points is then obtained by extending the shortest arc between them around the sphere. This process always yields a circle dividing the sphere into two equal hemispheres (for example the equator is a line in this sense).

From this description of lines, as circles on the surface of the earth whose centers are the center of the earth, it is easy to prove that in the geometry on the sphere, any two different lines will intersect. In particular, Euclid's parallel postulate does not hold for this geometric system, and so, as the large triangles drawn on the earth illustrated, neither does the sum-of-the-angles-in-a-triangle result. Note also that one consequence of what we know about the geometry on the surface of a sphere is that de Maria's two chalk marks on the floor of the Mojave Desert either have to intersect, if they are extended sufficiently far, or, if they are to remain twelve feet apart cannot both be lines.

Gauss understood that when a triangle is drawn on a curved surface, its angles could sum to a value other than 180 degrees. This is why Gauss designed his experiment as he did. Although the mountains themselves are on the surface of the earth, the triangle formed by their peaks is suspended in space—free of the earth's curvature. Yet, having freed his triangle from the curvature of the earth, Gauss faced a technical difficulty: How to determine the sides of an airborne triangle whose vertices are miles apart. To accomplish this Gauss appealed to an entirely reasonable misconception about light—that it travels in a straight line.

THE GEOMETRY OF SPACE

Escher's print *Smaller and Smaller* (Plate 3.2), provides an artistic representation of infinite divisibility, and Escher's print *Circle Limit III* (below) appears to do the same thing.

PLATE 8.1. *Circle Limit III*, 1959. M. C. Escher (1898–1972).
© 2007 The M. C. Escher Company-Holland. All rights reserved.

But there are important differences between the mathematical ideas underlying these two prints, which Escher alluded to in a lecture in 1964: "Instead of finishing with a square limit, one can also, and perhaps better, end with a circular outline. But this is no easy question, but a complicated, non-Euclidean problem."[2]

Smaller and Smaller is based on the same mathematical idea as Zeno's paradox, that it is possible to divide a magnitude in half, then in half again, repeating this division indefinitely. The images in *Circle Limit III* become smaller and smaller, moving from the center to the edges of the print, but their sizes are determined not through successive halving but by using non-Euclidean geometry. To make any sense of this last sentence it is necessary to reexamine what it means for a curve to be straight, and this reconsideration is aided by a closer look at Gauss's experiment.

Gauss's mountaintops were 43, 53, and 123 miles apart, so he could not apply Archimedes' conception of a straight line to determine his triangle's sides. Instead, Gauss appealed to a physical assumption based on Olaf Roemer's discovery in 1675 that light is not instantaneous, *the principle of least time*:

> When light travels from one point to another, it always
> follows the path requiring the least time.

Combining the principle of least time with Archimedes' conception of a line, Gauss assumed that light travels in a straight line. So, when Gauss looked from one vertex to another, his line of sight was along a line segment and so along a side of the triangle. Unfortunately, there are two flaws in such simplistic reasoning—one, which Gauss could accommodate, is the basis of the geometry in Escher's print, and the other, which Gauss could not have imagined, is a consequence of Albert Einstein's general theory of relativity.

It is possible to understand the non-Euclidean geometry Escher referred to above through an examination of one important consequence of the principle of least time. The seventeenth-century mathematician Pierre Fermat (1601–65) first espoused this important principle, not to explain why light travels in a straight line but why it does not. A straight stick, partly in air and partly in water appears to bend, and Fermat knew the cause of this illusion is the bending of the light traveling from the submerged portion of the stick to the observer's eye. Fermat used his principle to precisely calculate the path of light traveling from a medium of one density into a medium of a different density: How much the path of light bends is determined by the relationship between the two densities.

Suppose light travels from a point B, in water, to a point A, in air. There are many possible paths light could travel from B to A. It could travel along the straight line from B to A, or, in order to spend as little time in the water as possible, where it travels more slowly than in air, it could follow the dashed path below:

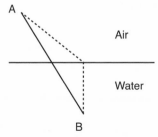

FIGURE 8.1. Two possible paths light could travel when traveling from a point B, in water, to a point A, in air. Fermat showed mathematically that the path light follows in moving from B to A lies somewhere between these two.

Before applying the refraction of light to provide a basis for the geometry of Escher's print, let's first reconsider the apparent bending of a stick when it is partly in air and partly in water. In Figure 8.2 a ray of light traveling from the end of the stick to the viewer's eye (at V) refracts and follows the indicated path. What the viewer sees is the image of the end of the stick at the point where the ray of light exits the water, so the end of the stick will appear to be at C.

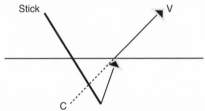

FIGURE 8.2. How the refraction of light makes a stick that is partly in water appear to be bent.

The arcs in Escher's print do not contain any sharp corners, so it is not immediately clear what role the refraction of light plays in their geometry. Before exploring this connection, let's see how refraction can lead to spectacular sunsets over the ocean. Because the density of the earth's atmosphere is greater at the surface of the earth than above it, a ray of light striking the atmosphere at an angle will be refracted (and so bent) as it moves from a region with a lesser density into one with a greater density. Nicole Oresme (c. 1320–82) appears to have been the first to realize that light could be continuously refracted as it passes through a medium with a uniformly varying density.[3] This means that at sunset, although the sun is below the horizon, it is still visible. The sunset lasts longer than the viewer expects because when the sun moves the relatively great distance from position A to its indicated position below the horizon, it seems to move only the shorter distance from A to its apparent position—the sun takes too much time to finally disappear below the horizon.

A

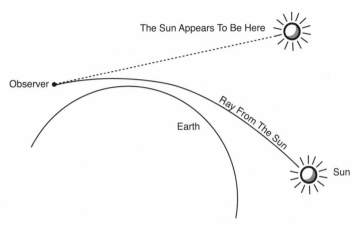

FIGURE 8.3. How the continuous refraction of light as it passes through the atmosphere can cause a sunset to last longer than expected.

Underlying Fermat's explanation for the bending of light is the assumption that although light is not instantaneous, its velocity through a medium of uniform density is constant—the denser the medium the slower light passing through it. Light refracts when it abruptly changes its velocity. A ray of sunlight at sunset is bent many, many times and follows a path that is not exactly curved but consists of many short, straight segments. If these segments are short enough, light's path appears to be a smooth curve (see Figure 8.3). This is the idea Escher needed.

Imagine you live inside Escher's disc (you are admittedly very flat, but play along), and that in this world, the path of a ray of light is taken to be a line segment. However, and you have no way of knowing this, the speed of light is not constant because the density of the space in the disc continually increases toward the disc's boundary. This property of light gives straight lines an unusual shape (although they will still be *straight* to you). To see what a line looks like to an observer from above the disc you can perform an experiment.

Suppose you are at point A, below, and you shine a flashlight toward point B. Viewed from above your beam of light will follow one of the three paths indicated in Figure 8.4.

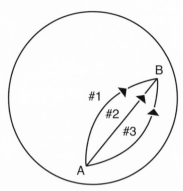

FIGURE 8.4. Three possible paths light could follow when traveling from point *A* to point *B* in a non-Euclidean disc.

Taking into account the behavior of light inside this disc, it is not too difficult to determine which of these paths is the correct one. Path no. 3 can be most easily excluded. If a ray of light were to travel along path no. 3 it would have to go farther than along path no. 2 and would travel slower than if it were on path no. 2 (because path no. 3 travels closer to the edge of the disc where light travels more slowly). So path no. 2 is more likely than path no. 3. What leads to the arcs in Escher's woodcut (and you have to do a rather serious calculation to discover this) is that light traveling along path no. 1 will get from A to B faster than light traveling along path no. 2. Even though path no. 1 is longer than path no. 2, along the middle portion of path no. 1, when the light is farthest from the edge of the disc, light is traveling so much faster than along path no. 2 that the trip along path no. 1 takes less time. Geometrically, a line segment through two points, in this world, is part of the circle that passes through these two points and meets the edge of the disc at two right angles.

Escher referred to this geometry as being non-Euclidean, and he did so because in this world Euclid's parallel postulate does not hold. In Figure 8.5 the line *L* crosses the other two lines at less than 90-degree angles, yet the two lines never intersect. Since the parallel postulate is not true, then neither is the result that the sum of the angles in any triangle equals 180 degrees. (In this non-Euclidean world the angles in any triangle will add to less than 180 degrees.)

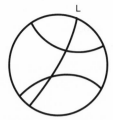

FIGURE 8.5. Why the parallel postulate does not hold for lines in a non-Euclidean disc.

When Gauss measured the angles between the rays of light forming his triangle he did not need to account for the refraction of light because the density of the air between the mountain peaks was more or less uniform. What Gauss could not account for was Einstein's discovery. According to general relativity, space would be flat, that is, it would satisfy the postulates of Euclidean geometry, if it did not contain any matter or energy. But matter and energy curve space; space is not a static medium containing the planets and stars, but a plastic one bent by massive bodies. It is possible that two lines, rays of light, can appear to be parallel, and even be parallel according to Euclid's parallel postulate, but after traveling a few miles, or light-years, bend and cross or merge into a single line and forever travel toward the universe's edge. Alternatively, in some triangles the angles sum to more than 180 degrees and in some to less than 180 degrees—this total does not depend on the triangle but on its location.

John Wheeler, one of the twentieth-century's leading physicists, provided an alternative perspective on the curvature of space: "There is nothing in the world except empty curved space. Matter, charge, electromagnetism, and other fields are only manifestations of the bending of space. *Physics is geometry*."[4] In Einstein's theory, bodies curve space; in Wheeler's theory, curved space produces bodies. Like the contrasting Renaissance views of matter—one, that earth, air, fire, and water have qualities that determine their chemical properties, and the other, that qualities combine to form earth, air, fire, and water—Einstein's and Wheeler's positions are complementary. Either the curvature of space or the existence of material bodies can be thought to precede the other; so on this point Pythagoras was right—geometry and the material cosmos are inseparable.

THE FOURTH DIMENSION

Picasso's *Les Demoiselles d'Avignon*, below, is frequently cited as the first modern painting. *Les Demoiselles d'Avignon* looks as if it could have been composed by first painting the five women, with their primitive masklike faces, onto a piece of glass, then shattering the glass and attempting to reconstruct the original painting from the salvageable pieces. But there is more to Picasso's painting than that, and to comprehend what Picasso achieved it is helpful to compare it with Cézanne's *Still Life with Fruit Basket* (Plate 7.6). Cézanne's still life offers a more or less straightforward representation of each of the objects on the table; it is just that different objects are represented as if viewed from different positions. *Les Demoiselles d'Avignon* is truly multidimensional; the components of the painting could not be rearranged to yield a convincing three-dimensional image because Picasso has painted his figures not just from multiple perspectives but as if they are being viewed through different lenses. Each figure has more than three dimensions; Picasso reveals qualities of each figure that are beyond our perception; qualities that are hidden from ordinary view. As likely a candidate as *Les Demoiselles d'Avignon* is for being the first modern painting, it also marks Picasso's movement toward cubism.

PLATE 8.2. *Les Demoiselles d'Avignon*, Paris, June-July 1907. Pablo Picasso (1881–1973). Oil on canvas, 8' × 7'8". Acquired through the Lille P. Bliss Bequest (333.1939). The Museum of Modern Art, New York. Digital image © The Museum of Modern Art / Licensed by Scala / Art Resource, New York. © 2007 Estate of Pablo Picasso / Artists Rights Society (ARS), New York.

In their book *Cubism and Culture* Mark Antliff and Patricia Leighton describe parallels between the prose style of Gertrude Stein (1874–1946) and cubist paintings (specifically Braque's *Violin and Palette* [1909] and Picasso's *Portrait of Wilhelm Uhde* [1910]).[5] While Braque and Picasso provide multidimensional views of their subjects, Stein uses nuance in language to offer multiple points of view. The difference, of course, is that the experience of reading unfolds in sequential time. The reader moves from sentence to sentence, as in the first few lines of Stein's 1912 description of Picasso: "One whom some were certainly following was one who was completely charming. One whom some were certainly following was one who was charming. One whom some were following was one who was completely charming. One whom some were following was one who was certainly completely charming."[6] When viewing a painting, the viewer's eyes are not necessarily drawn to the elements of the painting in the order the painter may have intended. A writer has more control. In Stein's few lines above, she sequentially invokes subtle changes in wording to evoke a multidimensional understanding of Picasso's charisma.

A more dramatic example of a multidimensional written image is the contemporary writer Robert Coover's short story "The Babysitter" (1969). This story is told through a series of paragraph-long descriptions of a teenage girl's evening caring for small children. The disconnected descriptions are told from the point of view of the girl, one of the children, the father of the children, and a couple of teenage boys. Each paragraph reports both the inner thoughts of one or more characters and their version of the actions of the girl. Like Picasso's *Les Demoiselles d'Avignon*, it is not possible to cut and paste Coover's story into a traditional narrative; it is not even possible to discern which actions are real and which are not. "The Babysitter" offers a cubist version of one evening—just as its characters' realities are combinations of both external reality and their own private thoughts, the reader's understanding of the story's reality is shaped (and shattered) by those combinations.

"The Babysitter" illustrates an essential assumption of the cubist approach to art or literature—there is more to material reality than can be captured through a photograph or narrative. For the visual artist, this means that objects have attributes, maybe even physical ones, beyond our immediate perception. The artist Max Weber (1881–1961) offered a geometric explanation for this discrepancy between perception, or representation, and reality (and in the process appealed to what was described in chapter 3 as being poetic infinity): "Two objects may be of like measurements, yet not appear to be of the same size, not because of some optical illusion, but because of a greater or lesser perception of [the] fourth dimension, the dimension of infinity."[7] This could be taken as the cubist manifesto. Cubists are not simply attempting to represent the psychological dimension that greatly influenced both literature and art following the work of Freud; this additional dimensionality is physical.

It is fairly easy to comprehend what it means to say that an object has a certain physical dimension, and it is natural to begin with a discussion of objects that are zero-dimensional. Elementary geometry trains us to say a point is zero-dimensional, but that does not assist us in understanding dimensionality. The simplest way to understand the concept of dimension is to first imagine that there can be a point without size, a line segment without width, and a square without height. Given these assumptions, consider what it means to say that the interior of a square, or a triangle, or a circle is two-dimensional. If these different-looking objects are to share the property of being two-dimensional, that property cannot be intrinsic to their shapes. To visualize this property, imagine each figure, a square, triangle, or circle, not as a drawing on a piece of paper, or a bent piece of wire, but as an infinitesimally thin object (a square can be visualized as a postage stamp without thickness, a circle as a flat coin).

The dimensions of an object can be understood by imagining the world of a being living inside it. Someone living inside a square, say at point P below, can clearly move from P to any other point, Q, in the square while staying within the square. To fix our ideas, imagine that our being travels from P to Q by moving to the right then up, or by moving up then to the right.

FIGURE 8.6. Two possible paths from P to Q in a two-dimensional square.

"Up" and "to the right" are independent because no amount of movement in one of these directions will in any way involve movement in the other. (There is another, shorter path, from P to Q, the diagonal path. But the diagonal direction is not independent of the up and right directions because a diagonal trip from P to Q can be achieved by combining the two.) You can get from any point in the square to any other point in the square by a sequence of up/down and right/left moves. (It is important to realize that up and down are the same directions of motion, if we allow for walking backward, and so think of moving down as moving up in a negative direction.) The world inside a square is a small two-dimensional space because it is possible to move from any position to any other position through a combination of the two independent directions, and so the square is a two-dimensional object.

This is not Weber's point of view. In Weber's conception of dimension, the square is not two-dimensional because a creature living inside it can move in either of two independent directions. Rather, a square is two-dimensional because it has two independent types of existence—it has a length (right/left nature) and a width (up/down nature). This language also suffices to describe what it means for an object to have three dimensions; it has an additional mode of existence (perhaps height), but fails to explain what it means for an object to be four-dimensional.

Although a mathematician would probably not embrace Weber's mysticism to explain how an object could be four-dimensional, Weber nonetheless helps us imagine how there could be another, concealed, fourth dimension. Imagine that space has not only three spatial dimensions but also a dimension of color, and that space can be any color of the continuous spectrum. Also imagine that spaces of all different colors exist simultaneously. Imagine further that just as an object can change its location it can change its color—when it takes on a new hue it is immediately transported to the space of that hue. The one catch in this world is that if an object has moved into a certain color world, such as the blue world, then it can only perceive or be perceived by other blue entities. A person living in the blue world would not be aware of any other colored world; if two blue people are standing side by side and one of them moves ever so slightly away from a pure blue hue, he will disappear.

The physical analogy to have in mind is that of a creature restrained to life in a two-dimensional plane, as in Edwin Abbott's book *Flatland: A Romance of Many Dimensions* (1884). These entities would have no awareness of the third, up-down, dimension. If one of the creatures in this world were to levitate above the plane, ever so slightly, he would suddenly become invisible to everyone else. Even though the levitated creature could be within a few three-dimensional inches of another creature, the plane-based creatures would have no awareness of the other one.

But this levitated creature would be completely aware of any other similarly levitated creatures. There is an entire two-dimensional world situated a few inches above the original one. And since these creatures can rise any distance above the original two-dimensional world, there will be a continuum of two-dimensional worlds stacked one upon the other. Taken together, all of these two-dimensional worlds constitute a world in which the creatures can move not only forward/backward, or left/right, but also up/down—taken together this continuum of two-dimensional worlds forms a three-dimensional world.

This description of dimension, which can also be used to establish a connection between zero- and one-dimensional spaces, or one- and two-dimensional objects, was alluded to by Duchamp in a discussion of the theory underlying his much-discussed piece *The Bride Stripped Bare by Her Bachelors, Even (The Large Glass)* (1915–23): This method involves "the repetition of a line . . . in order to generate the surface. [The same idea explains] passing from plane to volume [or from] the n-dim'l continuum to form the $n + 1$ dim'l continuum."[8] There are two ways to interpret Duchamp's explanation for how a series of one-dimensional lines can yield a two-dimensional surface, and these reflect two different assumptions about the nature of the continuum. (These are examined in chapter 10.) If a continuum is thought to consist of indivisibles, then placing a series of lines parallel to each other can form a surface (Figure 8.7, *left*). (The apparent measurable gaps between the vertical lines in this drawing are a consequence of our inability to represent adjacent indivisibles.) Alternatively, as illustrated on the right, if a continuum is seen as being infinitely divisible, then the lateral movement of a line can sweep out a surface.

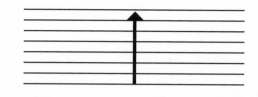

FIGURE 8.7. Two ways in which a line can be used to produce a plane. If the line consists of indivisibles (*left*), then drawing a line perpendicular to each indivisible produces a plane. Alternatively, if a line is infinitely divisible a plane can be swept out by moving the line in a lateral direction.

Applied to objects, this second approach reveals how a point can be moved to trace (and so form) a line segment, a line segment can be moved to form a square, and a square can be moved to form a cube. And although moving the cube in a new direction to form a four-dimensional hypercube cannot be visualized, it can be understood by analogy.

The American poet and impresario Walter Arensberg (1878–1954) described this last connection between dimensions, but in reverse, in his poem "Arithmetical Progression of the Verb 'To Be'" (1917):

> On a sheet of paper
>   dropped with the intention of demolishing
>   space
>   by the simple subtraction of a necessary plane
> draw a line that leaves the present
>   in addition
>   carrying forward to the uncounted columns
>   of the spatial ruin
>   now considered as complete
>   the remainder of the past.
>
> The act of disappearing
>   which in the three-dimensional
>   is the fate of the convergent
>   vista
>
> is thus
>   under the form of the immediate

arrested in a perfect parallel
   of being
   in part.[9]

While cubists may have attempted to represent the fourth dimension as a hidden physical reality, other artists sought to capture the fourth dimension through expressing feelings and emotions over objects and things. In his discussion of the fourth dimension, Weber also made this connection, "[The fourth dimension] is somewhat similar to color and depth in musical sounds. It arouses imagination and stirs emotion. It is the immensity of all things."[10]

The Russian painter Kazimir Malevich (1878–1935), in explaining the central tenet of his approach to painting, used the same language as Weber: "[The] appropriate means of representation is always the one which gives fullest possible expression to feeling as such and which ignores the familiar appearance of objects." Malevich then explains how in his "desperate attempt to free art from the ballast of objectivity" he began to paint pictures consisting of a black square on a white background: "The black square on the white field was the first form in which nonobjective feeling came to be expressed. The square = feeling, the white field = the void beyond this feeling."[11] This artistic philosophy has an earlier poetic analogue. Pure poetry (*poésie pure*) is so called not because it sought to represent Platonic ideals or otherworldly truths, but because it aspired to provide through language the same sensation as music. Edgar Allan Poe (1809–49), according to whom poetry is distinguished from prose in its appeal to lyricism over objectification, first enunciated the concept of pure poetry. The nineteenth-century French symbolist poets Baudelaire, Mallarmé, and Valéry incorporated Poe's idea into an entire poetic theory that sought to bring poetry closer to music and so closer to representing other dimensions of reality than the three of our senses. Poe's lyricism can be seen in the opening stanza of "The Raven" (1845):

Once upon a midnight dreary, while I pondered, weak and weary,
Over many a quaint and curious volume of forgotten lore,
While I nodded, nearly napping, suddenly there came a tapping,
As of some one gently rapping, rapping at my chamber door.

"'Tis some visitor," I muttered, "tapping at my chamber door—
Only this and nothing more."

PLASTIC TIME

[He] let his gaze wander to the swirling water of the stream racing madly beneath his feet. A piece of dancing driftwood caught his attention and his eyes followed it down the current. How slowly it appeared to move! What a sluggish stream!
— Ambrose Bierce, "An Occurrence at Owl Creek Bridge" (1891)

The illusion that the sun moves more and more slowly as it sets toward the horizon, or the companion observation that at sunrise the sun first moves very slowly then accelerates to its expected pace, illustrates the independence of experienced time from measured time. Writers were among the first artists to explore this independence. In the American writer Ambrose Bierce's (1842–c. 1914) story "An Occurrence at Owl Creek Bridge," both the overall narrative and individual passages describe a man's shifting experiences of time. At the beginning of the story, the man is about to be executed by being hung from a bridge over Owl Creek; the noose is already around his neck when he notices "the stream racing madly beneath his feet." Moments later it is "a sluggish stream." The man is dropped from the bridge, and time slows as the rope tightens on his neck; he is aware of pain shooting through his limbs and of increasing pressure in his skull. Suddenly the man hears a loud noise and plunges into the water below. The current moves him away from the soldiers on the bridge. The man eventually swims ashore, and after two days arrives back at his plantation. Just as he is about to embrace his wife: "a blinding white light blazes all about him, with a sound like the shock of a cannon—then all is darkness and silence!" The rope has pulled taut and snapped the man's neck.

James Joyce (1882–1941) went farther than Bierce in involving the element of time in literature. In Joyce's first novel, *A Portrait of the Artist as a Young Man* (1916), the passage of time is integrated into the style of writing. The novel begins: "Once upon a time and a very good time it was there was a moocow coming down along the road and this moocow that was coming down along the road met a nicens little boy named baby tuckoo." The grammar, diction, and complexity of the writing mirror the narrator's intellectual development from childlike to scholastic. The reader is propelled through the time of the novel by both the narrative and its style.

In *Ulysses* (1922) Joyce employs time in an entirely different manner; time is expanded both in the structure of the narration and in particular passages. The narration chronicles Leopold Bloom's movement from encounter to encounter, and from pub to pub, throughout a single day (June 16, 1904). The day, and the novel, ends with Molly Bloom's notorious soliloquy. Although time moves from the past to the future, its pace is not constant.

Finnegans Wake (1939) is Joyce's last, and most ambitious, novel. It begins with the sentence fragment "riverrun, past Eve and Adam's, from swerve of shore to bend of bay, brings us by a commodius vicus of recirculation back to Howth Castle and Environs" and ends with a sentence fragment that can be attached to the beginning of the opening sentence fragment: "A way a lone a last a loved a long the[.]" The novel's structure is cyclic; the narration does not move from the past to the future. The reader could start anywhere in the novel.

It took visual artists longer than writers to present time as an independent element in their art. While medieval and Renaissance artists employed continuous narrative, and thereby incorporated various views of the same story within a fixed space (e.g., Gozzoli's *Arrival of St. Augustine in Milan*, Plate 7.4), they did not portray time in their paintings. Their use of multiple scenes was simply a convention allowing the painter to tell a story without executing several paintings. By the twentieth century, artists had begun to attempt to represent time. Instead of using multiple points of view for a single scene, or multiple representations of some chronology, artists sought to represent or draw our attention to time. Two early examples of this, both from 1912, are Duchamp's *Nude Descending a Staircase (No. 2)* (Plate 3.3) and Giorgio de Chirico's *Enigma of the Hour*, below.

Duchamp and de Chirico (1888–1978) used time in entirely different ways. Duchamp described his painting as being a study of how to represent movement. The figure is not presented to illustrate different scenes from a single narrative, as in the use of continuous narrative, but to illustrate a single act—the movement of the body through space.

PLATE 8.3. *Enigma of the Hour*, 1912. Giorgio de Chirico (1888–1978). Mattioli Collection, Milan, Italy. Photo: Scala / Art Resource, New York. © 2007 Artists Rights Society (ARS), New York / SIAE, Rome.

Duchamp's painting can almost be said to illustrate duration, the span of time required for any physical act. De Chirico's goal was different; he sought to undermine our confidence that we can understand the relationship between the perception of time and the reality of time. In *Enigma of the Hour* physical time and measured time are in conflict. Although the clock reads 2:54, presumably in the afternoon, the shadows indicate that it is either sunrise or sunset. Time is an independent dimension—bound neither to our expectations nor to physical space.

In his special theory of relativity (1905) Einstein proved that time's independence from experience is not just psychological. If one thousand clocks were all set to the same time, dispersed across the universe to various comets and planets, and then compared after a few earth days, none of them would very likely agree. The faster the velocity of the clock over the few days it was away from earth, the slower it would have run. A clock moving near the speed of light would have ticked off only a few seconds—a person accompanying the clock would hardly have aged at all.

POSTSCRIPT: GEOMETRY WITHOUT ANGLES

The result that the sum of the angles in any triangle equals 180 degrees was used to develop Plato's theory of matter, because it follows from this result that there are exactly five Platonic solids. However, there is another way to discover these solids that has nothing to do with the measurement of angles; it has more to do with examining the properties of geometric objects that remain if the object is stretched but not torn. Over the past 150 years, mathematicians have developed this important type of geometry, known as *topology*. It has been invaluable to mathematical research since its inception.

To understand how topological considerations can lead us back to the five Platonic solids, consider, for example, an ordinary cube that has been deformed slightly. This new object retains some properties of the cube, but not others. In particular, a slightly deformed cube still has six sides, and each side has four edges, even though the sides are no longer perfect squares. Also, three edges will still meet at each corner of the deformed cube. The defining property of this object is not that its sides are squares, but that its edges have the properties described in the previous sentences; so, we will view this deformed cube not as a solid, but as a collection of twelve edges (as if our original cube had been made out of pipe cleaners instead of solid squares). Before we consider the connected edges of the other Platonic solids, we examine a simpler object—a doodle on a piece of paper.

We begin with what we mean by a doodle. To form a doodle we can start with a point (a vertex) inside a rectangle and add arcs. The only rules are that every arc must begin and end with a vertex, and if two arcs cross there is a vertex where they cross. The two possible outcomes of this process of beginning with the single point P and adding arcs, are given by doodle 1 and doodle 2, below. Doodle 3 and doodle 4 represent doodles constructed by literally "doodling" on a piece of paper and then putting a vertex wherever two lines cross and at the end of any dangling lines.

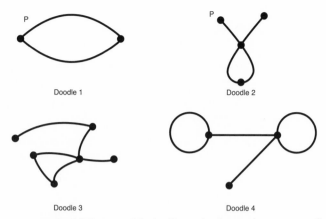

FIGURE 8.8. Four possible doodles with edges and vertices.

Each of the above figures consists of points (vertices) and arcs (edges), and what is less immediate but apparent once it is pointed out is that each figure subdivides the interior of the rectangle into regions. For example doodle 2 consists of four vertices and four edges, and separates the rectangle into two regions. The information for each of the figures in Figure 8.8 is given below.

	Edges	Vertices	Regions
doodle 1	2	2	2
doodle 2	4	4	2
doodle 3	6	6	2
doodle 4	4	3	3

There is a simple, and unexpected, formula relating the number of edges, E, and the number of vertices, V, of the doodle, and the number of regions, R, into which the doodle subdivides the rectangle. Leonhard Euler (1707–83) discovered this formula in the eighteenth century, and it says that for any doodle inside a rectangle

$$R + V = E + 2$$

This amazing formula does not just apply to doodles inside a rectangle, provided we count the number of regions the doodle defines properly. If we imagine that our figure is drawn on any surface, for example on a sphere, then the formula still holds as long as we remember to count the "outside" region (for example a triangle on the surface of a sphere defines two regions—an inside and an outside).

The reason this formula is correct is easy to grasp. We visualize constructing a doodle from the simplest ones by adding edges and vertices. The most basic doodle is a single point without any edges. This doodle has one vertex, no edges, and defines one region (all of it being "outside"). Thus Euler's formula holds because when we plug in the appropriate numbers, $R + V = E + 2$ becomes $1 + 1 = 0 + 2$. Imagine how we can form a more complicated doodle from the simple dot: We can add an edge that loops back to the single vertex (doodle 5, below), or we can add an edge that has our original vertex on one end, and a new vertex on the other, recalling that every edge must begin and end at a vertex (doodle 6, below).

Doodle 5   Doodle 6

FIGURE 8.9. The two possible doodles that can be formed by adding a single edge to a doodle consisting of a single vertex but no edge.

When we go from the original dot to doodle 5, we have added one edge and one region. Since the relationship $R + V = E + 2$ holds for the original doodle, it must hold for the doodle in doodle 5, since both R and E are increased by one (so we obtain the equation for the doodle in doodle 5 by adding one to each side of the equation for the original doodle). Similarly, when we go from the original doodle to the one in doodle 6 we have added one edge and one vertex, so beginning with the equation $R + V = E + 2$, which holds for the original doodle, we see that it must also hold for doodle 6 since we are just adding one to each side. As any doodle can be constructed by beginning with a single vertex and then adding edges one at a time, Euler's formula holds for all doodles since it holds for a single point.

It is possible to rediscover the five Platonic solids using Euler's formula. The crucial step is to visualize a Platonic solid as a doodle on the surface of a sphere. What is needed is a complete determination of all doodles on the sphere that satisfy

1. every region is enclosed by the same number of edges, and
2. the same number of edges meet at every vertex.

It is important to observe that for a doodle to represent a solid made up of polygons, every region must be enclosed by at least three edges, and at least three edges must meet at every vertex. Assuming these restrictions, it is not too hard a calculation to discover that there are only five doodles on a sphere satisfying the two conditions, above. These five doodles correspond to the five Platonic solids.

What Is a Number?

9

There is an old and a new consciousness of time.
The old is connected with the individual.
The new is connected with the universal.
The struggle of the individual against the universal is
  revealing itself in . . . the art of the present day.
  — De Stijl (1918)

In 1917 the artist Theo van Doesburg (1883–1931) published the first edition of *De Stijl* [The Style], a journal dedicated to promoting aesthetic values based on isolating, and then representing, the basic geometric components of art and architecture. The first number of the second volume contained "Manifest 1 . . . 1918," in Dutch, French, English, and German. The manifesto began with the proclamation above; the third proclamation read: "The new art has brought forward . . . a balance between the universal and the individual." This manifesto concluded with the signatures of a group of artists, not all of whom had met.

Signatures of the present collaborators: Antony Kok, *Poet*
Theo Van Doesburg, *Painter*   Piet Mondriaan, *Painter*
Robt. Van 'T Hoff, *Architect*   G. Vantongerloo, *Sculptor*
Vilmos Huszar, *Painter*   Jan Wils, *Architect*[1]

One manifestation of this "balance between the universal and the individual" that the members of De Stijl sought was art and architecture based on the mathematical principle of orthogonality—vertical and horizontal elements meeting at 90-degree angles. About the time "Manifest I" appeared above their signatures, both van Doesburg and Piet Mondrian (1872–1944) were basing their paintings on grids of vertical and horizontal lines. However, there were subtle, and eventually irreconcilable, differences between the way in which these two artists conceived of their art; to understand this it is important to understand the different ways they employed the grid in their paintings.

Mondrian's paintings consisted of a black grid of intersecting lines that subdivided the canvas into rectangles of various sizes. Mondrian painted each of these rectangles a single color, either a primary color, black, white, or grey. In van Doesburg's paintings the grid was not defined by black, intersecting horizontal and vertical lines. Rather the grid was represented by the spaces between painted rectangles (planes)— the colored rectangles did not have borders.

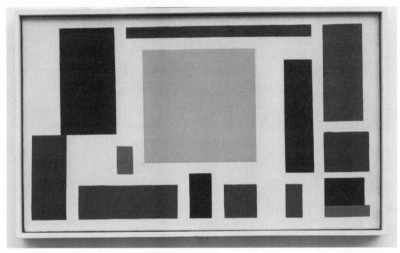

PLATE 9.1. *Composition VIII (The Cow)*, c. 1918. Theo van Doesburg (1883–1931). Oil on canvas, 14¾ × 25 in. Purchase (225.1948). The Museum of Modern Art, New York. Digital image © The Museum of Modern Art/Licensed by Scala / Art Resource, New York.

This might not seem like much of a difference but it implies that van Doesburg saw the grid as a tool for expressing relationships between the colored planes, while, for Mondrian, the grid was an element of the painting. Mondrian sought harmony through the geometry of the painting and through the use of primary colors. Van Doesburg maintained that such harmony could only be achieved through "colors far removed from each other, colors of unequal value, contrasting colors, dissonants, achieving a unity through the relationship from color to color."[2]

Van Doesburg further drifted from the use of the rectangular grid in 1925 when he displayed paintings with diagonal as well as vertical and horizontal elements.

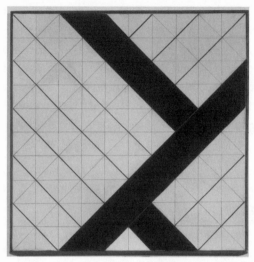

PLATE 9.2. *Counter-Composition VI*, 1925. Theo van Doesburg (1883–1931). Oil on canvas, 50 × 50 cm. Tate Gallery: London. Photo: Tate, London / Art Resource, New York.

Of his use of the diagonal, van Doesburg wrote: "By assuming a new direction in relation to the direction[s] already known ... [we] make our consciousness accessible to a new polarity."[3] The official split between Mondrian and van Doesburg occurred in the spring of 1925—not so much over van Doesburg's introduction of diagonals as much as over his insistence that the diagonal was needed to achieve what Mondrian felt he had already achieved through the use of vertical and horizontal elements. (Mondrian himself had experimented with diagonal elements but not in combination with horizontal and vertical lines.)

Another signatory of the De Stijl manifesto was the painter and sculptor Georges Vantongerloo (1886–1965). The youngest member of De Stijl, Vantongerloo, at least initially, adhered more stringently to the vertical-horizontal principle than either Mondrian or van Doesburg, but by 1926 his conception of the role of geometry in his artwork was much more expansive. Vantongerloo sought what he called the "unity" of a piece of art. In his *Reflections III* (1926) he wrote: "The principle of unity consists in finding the elements of some geometrical form or algebraic equation and creating a new geometrical form which has the elementary form as the basis of its unity."[4] These "new geometric forms" arising from both geometry and algebra are the inspirations for two of Vantongerloo's pieces:

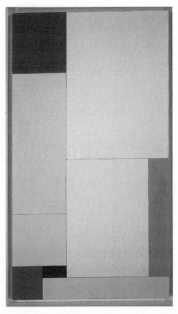

PLATE 9.3. *Composition Derived from the Equation $y=ax^2 + bx + 18$ with Green, Orange, Violet (Black)*, 1930. Georges Vantongerloo (1886–1965). Oil on canvas, 47 × 26⅞ in (119.4 × 68.2 cm). Solomon R. Guggenheim Museum, New York (51.1299). © 2007 Artist Rights Society (ARS), New York / ProLitteris, Zurich.

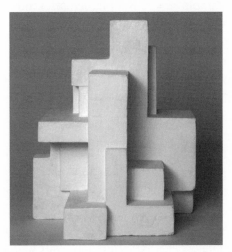

PLATE 9.4. *Construction of Volumetric Interrelationships Derived from the Inscribed Square and the Square Circumscribed by a Circle*, 1924. Georges Vantongerloo (1886–1965). Cement cast painted white. Height: 11¹³/₁₆ in (30 cm). The Solomon R. Guggenheim Foundation, Peggy Guggenheim Collection, 1976 (76.2553.59). © 2007 Artist Rights Society (ARS), New York / ProLitteris, Zurich.

In executing these two pieces Vantongerloo did not follow mathematical formulas or patterns. Vantongerloo wrote about his creative process, "Clearly the important thing here is to know how to create. Creation is not a formula but a principle.... As this principle has unity as a basis ... it is universal, immutable, infinite, eternal."[5] So rather than work with formulas, Vantongerloo sought to represent unity in his art, and as he wrote elsewhere, this unity could be found through the discovery of relationships. In Vantongerloo's painting *Composition Derived from the Equation* ... (Plate 9.3), the relationship appears to be between areas and color. In the sculpture (Plate 9.4) the relationship seems to be between volumes. What is not clear from just looking at these two pieces but is implicit in their titles is that behind each of these works lies a different answer to the question "what is a number?"

AN INTRODUCTION TO ALGEBRA

So far, we have only discussed mathematical ideas associated with geometry, or with whole numbers and their ratios, and the only equations that have appeared here, except for the one expressing the relationship given by the Pythagorean theorem, have been equalities of two ratios, for example $\frac{1}{2} = \frac{2}{4}$. But mathematics consists of more than just geometry and proportions, it includes among its many subdisciplines algebra, which concerns itself with the study of equations such as the one in the title of Vantongerloo's painting.

The history of algebra not only provides a framework within which the history of almost all mathematical ideals can be understood, but it is especially important in the development of the modern concept of number. Indeed, it is easier to see the evolution of the idea of number in attempts to find solutions to algebraic equations than in attempts to provide numerical equivalents for all geometric magnitudes—had it not been for the invention of algebra the notion of number would still be a very limited one.

In order to understand the algebraic concept of a number, it is useful to first understand how algebraic equations were understood in Euclidean geometry. In Greek mathematics, a simple equation such as $2 \times L = 10$ was not written symbolically, but it could have been given rhetorically. More importantly, in aiding our understanding of Vantongerloo's piece, such a relationship was viewed as a statement about areas. The unknown quantity L was thought to be the unknown length of the base of a rectangle, whose height is 2 and whose area equals 10 (since the area of a rectangle equals its height times its base: $area = 2 \times L = 10$).

Similarly, a more elaborate algebraic equation, such as $(a + b)^2 = a^2 + 2ab + b^2$, was also viewed as a statement about areas of squares and rectangles. This equation reflects the geometric result shown in Figure 9.1: the area of the large square equals the sum of the areas of the two smaller squares and the two smaller rectangles.

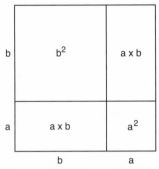

FIGURE 9.1. A geometric justification for the algebraic relationship $(a + b)^2 = a^2 + 2ab + b^2$. The area of the large, all-encompassing square is $(a + b)^2$ and the sum of the areas of the small, enclosed squares and rectangles is $a^2 + 2ab + b^2$.

The equality-of-areas interpretation of this equation is the one given in Euclid's *Elements*, and his statement of this result reveals how awkward it was to express algebraic identities without algebraic notation: "If a straight line is cut at random, the square on the whole is equal to the squares on the segments and twice the rectangle contained by the cut segments." This result says that two different areas, based on a given line, are equal. The first is the area of the square whose sides equal the length of the segment, the large square above. The second area is the sum of four areas, based on the pieces obtained when the segment is divided arbitrarily into two pieces. These four areas consist of a square whose sides are determined by one of the pieces, a square whose sides are determined by the other of the two pieces, and two copies of the rectangle formed by using one of the pieces for its height and one of the pieces for its base. These four areas, which are the figures inside the large square in Figure 9.1, can be recombined to make the large square.

Not until the third century A.D. did Greek mathematicians move from examining only algebraic identities involving geometric quantities to studying algebraic equations without regard for their geometric content. This shift in point of view is evident in the work of the mathematician Diophantus. Diophantus sought solutions to equations, involving one or more unknowns, where the solutions were whole numbers. Diophantus did allow for the use of fractions; he thought of them as parts of a whole and not as numbers existing independent of some whole. For example, ½ was not seen as being a number but as one-half of a whole.

Diophantus is also credited with introducing a symbolic notation for these equations. As an example, Diophantus discussed the then already-ancient problem of finding whole numbers X, Y, and Z that are solutions to the equations $X^2 + Y^2 = Z^2$, in other words whole numbers that can be the sides of a triangle with a right angle. (By the Pythagorean theorem, if X, Y, and Z are the sides of the triangle, where Z represents the side opposite the right angle, then they satisfy $X^2 + Y^2 = Z^2$ [see chapter 1].)

In the first millennium A.D., Arab and Hindu mathematicians made the greatest advances in algebra. The most relevant of these advances, to our first examination of Vantongerloo's painting, are the accomplishments of Arab mathematicians at the House of Wisdom (in Baghdad)—especially those of Abu Ja'far Muhammad ibn Musa al-Khwarizmi (c. 780–847). Al-Khwarizmi wrote two texts that greatly influenced the development of mathematics in medieval Europe. The second, *Treatise on Calculation with Hindu Numerals* (c. 825), showed how to represent counting numbers with the numerals 1 through 9, using 0 as a place holder (for example, 1066). This text also included procedures for calculating with these numerals, procedures that have come to be known by a title derived from al-Khwarizmi's name—algorithm.

The first text, *Compendious Book on Calculation by Completion and Balancing* (c. 820), showed how to solve certain algebraic equations. The Arab scholars in Baghdad had access to most Greek mathematics and they knew that the Greeks had understood algebraic identities and algebraic equations. One of al-Khwarizmi's greatest accomplishments was his thorough examination of quadratic equations. Al-Khwarizmi sought positive solutions to these equations, whether those solutions were whole numbers, rational numbers, or even irrational numbers. Having this as his goal, al-Khwarizmi determined which quadratic equations had positive solutions; he found these equations could be put into one of six forms:

1. $aX^2 = bX$
2. $aX^2 = c$
3. $bX = c$
4. $aX^2 + bX = c$
5. $aX^2 + c = bX$
6. $aX^2 = bX + c$

each of which is viewed as an equality of areas. For example, the equation $X^2 = 7$ would be interpreted as "find the side of a square, so that the area of the square is seven square units," while the equation $\frac{1}{2} X^2 = 7$ would be interpreted as "find the side of a square so that one-half of its area is seven square units." Stated in terms of equality of areas, each of these problems is one that the Greek mathematicians would have considered; the Greeks just would not have allowed the symbol $\frac{1}{2}$ in an equation. Notice that al-Khwarizmi did not consider the familiar equation (at least to us and to Vantongerloo) $aX^2 + bX + c = 0$, with positive a, b, and c, because this equation does not appear to be a statement about areas (three areas cannot be added to give an area equaling zero).

Al-Khwarizmi indicated how to find solutions to these equations, and his method for solving these equations is what connects Vantongerloo's painting to him. To see this connection, consider how al-Khwarizmi solved the equation $X^2 + 10X = 39$. Al-Khwarizmi's method is to begin with the left-hand side of the equation $X^2 + 10X$ and view it as a combination of two areas: X^2 is the area of a square whose sides are of length X, and $10X$ is the area of a rectangle whose length is 10 and height is X. The plus sign in the equation $X^2 + 10X = 39$ is an indication that these two areas should be combined, ideally into a square, whose total area is 39.

But two geometric figures, such as a rectangle and a square, cannot easily be merged into a single square. Following the ideas that were developed in Euclid's *Elements*, al-Khwarizmi combined these two areas by first imagining the 10 by X rectangle as two 5 by X rectangles:

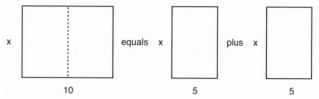

FIGURE 9.2. In order to solve algebraic equations geometrically, it is necessary to rearrange areas, for example, by decomposing the area of a single rectangle into the sum of the areas of two smaller rectangles.

Combining the area for X^2 with these two areas we almost obtain a square; it is missing the 5 by 5 square:

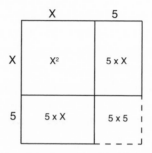

FIGURE 9.3. To solve the original equation $X^2 + 10X = 39$ we need the area of the combined areas of the three solidly outlined areas to equal 39.

We want the total area contained in the three solidly outlined regions to equal 39, so that the area of the entire square should be $39 + 25 = 64$. Thus we will have found a solution to the original equation if we can find a solution to the new equation: $X^2 + 5X + 5X + 25 = 39 + 25 = 64$, which is the same as $(X + 5)^2 = 64$. Since $8 \times 8 = 64$ the solution is $X + 5 = 8$ so $X = 3$.

This example was specifically chosen so that the numbers worked together to yield a final answer, $X = 3$, which is a whole number. But the geometry does not usually yield such a simple solution. If the original equation were changed very little, for example to $X^2 + 10X = 40$, the geometric technique employed above requires us to solve the equation $(X + 5)^2 = 40 + 25 = 65$. Taking the square root of both sides then gives the equation: $X + 5 = \sqrt{65}$, so $X = \sqrt{65} - 5$, which is an irrational, positive quantity.

ALGEBRAIC RELATIONSHIPS

At least superficially, Vantongerloo's painting *Composition Derived from the Equation $y = ax^2 + bx + 18$ with Green, Orange, Violet (Black)* reveals that, like the Greeks' and al-Khwarizmi's geometric method for finding a solution to an algebraic equation, Vantongerloo thought of his equation as a relationship between areas. So in this example perhaps the "new geometric form" Vantongerloo sought to create was nothing other than the ancient geometric solution to the algebraic equation $aX^2 + bX + 18 = 0$. Unfortunately, there is no evidence that Vantongerloo was thinking of the Greek and Arab geometric methods for solving algebraic equations, and he alludes to another way of viewing an equation such as $Y = aX^2 + bX + 18$ later in *Reflections III*: "The Greeks spoke to us of proportion. In the new art we speak of relations: relation within the work and in relation to Unity."[6] We can infer from this that Vantongerloo did not view an equation as being static. He understood that an equation such as $Y = aX^2 + bX + 18$ establishes a dynamic relationship between two entities, for example between X and Y.

To understand this shift from a static point of view to a dynamic point of view, consider the equation $Y = aX^2 + bX + 18$. When $X = 1$ the value of Y is determined (in terms of the coefficients a and b), and when $X = 2$ another value of Y is determined. These are two static pieces of information about the relationship $Y = aX^2 + bX + 18$, but Vantongerloo sought to understand not just the relationships between particular values of X and Y but the meaning of the relationship itself. This is precisely what René Descartes (1596–1650) had made possible with his invention of coordinate geometry. Descartes, of course, realized that the relationship between X and Y can be visualized by graphing X versus Y on the coordinate axes. Descartes' idea was simple, but it was also profound: To each algebraic relationship, or equation, it associates a geometric object. A simple relationship such as $Y = 2X + 1$ corresponds to a line, while a quadratic relationship such as $Y = X^2 - 1$ corresponds to a parabola. (Some of these geometric objects would have been recognizable to the Greeks. They had studied both lines and so-called conic sections, which include circles, ellipses, and parabolas, but a simple equation like $Y = X^4 + 1$ would have had no meaning to them because while X^2 represents the area of a square, interpreting the meaning of X^4 would require thinking in four dimensions.)

Vantongerloo sought to artistically interpret the relationship this equation establishes between X and Y by moving beyond Descartes' basic graphical representation. Vantongerloo wrote that he wanted to "reveal the incommensurable" in the geometric object underlying the equation. He wrote in his "Introductory Reflections": "The coordinates X and Y are imposed upon us solely because they are convenient and not because they are indispensable. All of this is too rigid for freedom. We must escape."[7] Vantongerloo escaped by reinterpreting the area relationships implicit in the algebraic relationship $Y = aX^2 + bX + 18$. We can obtain what might be a glimpse into Vantongerloo's thinking when we note that he provided a sculptural interpretation of the slightly different equation $Y = -aX^2 + bX + 18$ as a combination of rectangular blocks.

It is apparent that Vantongerloo saw other implicit relationships between X and Y. It is also apparent that Vantongerloo's painting provides us with an example of the creativity he referred to in his *Reflections III* rather than information about solutions to algebraic equations. Yet understanding solutions to algebraic equations is so important to understanding the modern conception of number that we briefly leave Vantongerloo and return to the history of algebra.

ALGEBRA REVISITED

The most important figure in bringing algebra to Europe was Leonardo of Pisa (c. 1170–1240), also known as Fibonacci. Fibonacci did not interpret all algebraic equations as geometric relationships; he went so far as to propose that irrational quantities be treated as numbers rather than geometric magnitudes. One of Fibonacci's greatest achievements was to show that even allowing geometric magnitudes involving square roots, such as $\sqrt{2}$ or $7 + \sqrt{28}$, to be numbers, there were still algebraic equations without solutions. Without explaining how he found it, Fibonacci produced a solution to the equation $X^3 + 2X^2 + 10X = 20$, and then showed that this number was not a number that can be expressed in the form of irrational numbers arising in geometry.

The first European mathematician to make significant contributions toward solving algebraic equations was Girolamo Cardano (1501–76). Cardano's *Ars Magna* (1545) illustrated the power of algebraic techniques for solving practical problems, and greatly extended the types of algebraic equations that could be solved (although he did not extend the concept of number). Cardano himself was quite impressed with his accomplishments, as the title page of the book indicates:

THE GREAT ART

OR

THE RULES OF ALGEBRA

BY GIROLAMO CARDANO

OUTSTANDING MATHEMATICIAN, PHILOSOPHER AND PHYSICIAN

The book begins: "In this book, learned reader, you have the rules of algebra. It is so replete with new discoveries, and demonstrations by the author—more than seventy of them—that its forerunners [are] of little account," and ends with

WRITTEN IN FIVE YEARS, MAY IT LAST

AS MANY THOUSANDS

THE END OF THE GREAT ART ON

THE RULES OF ALGEBRA

BY GIROLAMO CARDANO.

Between these two extravagances, which are more restrained in the original Latin, Cardano provided formulas for solving cubic and biquadratic equations, that is equations such as $X^3 + X = 6$ and $X^4 + 3X^2 = 4$. (A formula for solving "quadratic" equations had more or less been established by the Babylonians and transferred to Italy through the Arab mathematicians.)

The reason Cardano wrote in his introduction that the book involves more than seventy discoveries and demonstrations is that he considered all possible combinations of powers of the unknown, X, as had al-Khwarizmi. This permitted him to employ geometric arguments where appropriate. Even with these various forms for equations, Cardano was led to fictitious solutions, such as square roots of negative numbers. These may be seen in one of Cardano's sample problems, which seems innocent enough:

> Find two numbers that add up to 10 and
> give 40 when they are multiplied together.

If we denote these unknowns by X and Y, then our requirement that when we multiply them we get 40 can be expressed as the equation: $XY = 40$. The problem also asks that the numbers add to 10, so $X + Y = 10$. This last equation relates X and Y in a simple fashion: Solving it for Y we obtain $Y = 10 - X$. If we then substitute this expression for Y into the equation $XY = 40$, we obtain the equation: $X(10 - X) = 40$. The quadratic formula, which was understood by Cardano, says that the two numbers are: $X = 5 + \sqrt{-15}$ and $Y = 5 - \sqrt{-15}$. Although the square root of the negative fifteen was not something he considered to be a number, Cardano wrote, "Putting aside the mental tortures involved, multiply [these quantities]. Hence the product is 40." Cardano continued with what has become one of his most widely misquoted sentences, "So progresses arithmetic subtlety the end of which, as is said, is as refined as it is useless."[8]

This comment from Cardano should be contrasted with one in another important book on algebra published less than a century later, Albert Girard's *L'Invention nouvelle en l'algebra* (1629).[9] Concerning the square roots of negative numbers, Girard posed the rhetorical question, "Of what use are these impossible solutions?" He then answered that they are important for three things, "for the certitude of the general rules, for their utility, and because there are no other solutions."[10] In arguing for the acceptance of these numbers, Girard did not appeal directly to their beauty, he based his arguments on more practical considerations. But his last reason for their acceptance, "because there are no other solutions," is an aesthetic consideration. It is based on the mathematical notion of elegance, which is examined in chapter 12.

Mathematicians eventually took Girard's advice, but not until the nineteenth century, and adopted the following definition:

> A number is a solution to an algebraic equation.

Notice that this definition includes all the positive and negative counting numbers (for example -3 is the solution to $X + 3 = 0$) and all of the positive and negative rational numbers ($2/3$ is the solution to $3X - 2 = 0$) and so encompasses all of the most commonly encountered numbers.

But it also included the possibly troublesome $\sqrt{-1}$ and the solutions to Cardano's problem above.

A GEOMETRIC DEFINITION OF NUMBER

Number is that which expresseth the quantitie of each thing.
— *Simon Stevin, "Disme: The Art of Tenths" (1585)*

Our examination of Vantongerloo's painting *Composition Derived from the Equation y = ax² + bx + 18 with Green, Orange, Violet (Black)* and its superficial similarity to al-Khwarizmi's geometric solutions to his six types of quadratic equations led us into a discussion of algebra and to the definition of a number: A number is a solution to an algebraic equation. But there was another, competing definition of number, and Vantongerloo's sculpture *Construction of Volumetric Interrelationships Derived from the Inscribed Square and the Square Circumscribed by a Circle* is associated with its evolution. This is the concept of number deriving from geometric magnitudes.

In "Reflections III" Vantongerloo illustrated the geometry behind his sculpture by drawing a side view of the sculpture superimposed over a square and two concentric circles:

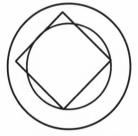

FIGURE 9.4. In Vantongerloo's drawing the sculpture is contained in the above square.

In Figure 9.4 the square is not centered on the circles but is partly inscribed within the outer circle and partly inscribed inside and circumscribed outside the inner circle. Vantongerloo's square is an artistic hybrid of Figure 9.5.

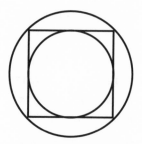

FIGURE 9.5. A square contained between two concentric circles.

This last drawing can be associated with one of the greatest mathematical accomplishments of antiquity, Archimedes' estimate for the geometric magnitude π.

We have already used the result from Greek geometry that whenever two triangles have the same shape the ratios of their corresponding sides are equal. Greek geometry had a similar result for circles: Given any two circles, the ratios of their circumferences and diameters are equal. So if one circle has circumference C_1, and diameter d_1, and another has circumference C_2, and diameter d_2, we have the following equality:

$$C_1/d_1 = C_2/d_2.$$

This is an expression for the equality of ratios of geometric magnitudes, and Greek mathematicians were perfectly comfortable accepting this equality as a universal truth without having to think of either of these ratios, say C_1/d_1, as a number. All the Greeks would have said is that the ratio of the circumference to diameter of a circle is independent of the circle. But for us, the relationship between the circumference and diameter of a circle is given by the formula

$$circumference = \pi \times diameter$$

so the ratios C_1/d_1 and C_2/d_2 are equal because they each equal the number π.

This formula was not known to Greek mathematicians, and could not have been, because it violates their conception of geometry and number. The very existence of our formula assumes that π is a number and that both the circumference and diameter of a circle, two geometric magnitudes, could be quantified by numbers. Put differently, for us the above formula is a simple—even beautiful—relationship between three numbers, and this formula may be freely used to find the diameter of a circle given its circumference or its circumference given its diameter.

One of Archimedes' many mathematical achievements was to obtain an estimate for the ratio of the circumference of a circle to its diameter—our π. Archimedes' estimate of π was based on two observations. The first is that given a circle with circumference C, it is possible to approximate C by the perimeter of a polygon inscribed inside, or circumscribed around, the circle:

FIGURE 9.6. In these drawings, the circumference of the circle is approximated by the perimeters of inscribed and circumscribed pentagons and hexagons. Approximating the circle by an inscribed and circumscribed square may have inspired Vantongerloo.

The second observation is that the more sides the polygon has, the better it approximates the circle. If you estimate the circumference of a circle by the perimeter of an inscribed polygon, and then by the perimeter of an inscribed polygon with twice as many sides, and then by the perimeter of an inscribed polygon with twice as many sides again, you will obtain better and better approximations of the circumference. There is no intrinsic limit as to how well the circumference may be approximated by perimeters of polygons; any obstacle to obtaining a very, very good estimate of the circumference emerges from the computational difficulty of the problem. (Although the perimeter of a polygon with a modest number of sides, such as four or eight or sixteen, is easy to compute, the calculation becomes more and more difficult as the number of sides increases.) The point is that although using this process you will never obtain the circumference, you will obtain better and better approximations of it. Archimedes carried out this calculation for inscribed, and circumscribed, polygons with ninety-six sides and, in modern notation, concluded that

$$3^{10}/_{71} < \pi < 3^{1}/_{7}.$$

As recently as the sixteenth century, π was not a considered to be a number but a geometric magnitude that could be estimated by the ratios of numbers. While negative and irrational numbers were only slowly accepted as numbers, there arose in surveying and astronomy the need for numbers to represent more accurate measurements. The standard way to represent, for example, the measurement of an angle, is to use degrees, minutes, and seconds (where sixty minutes equals one degree, $60' = 1°$, and sixty seconds equals one minute, $60'' = 1'$). But this method is both awkward and limited. If we want to add two angles together, such as $5°50'44''$ and $7°25'54''$, we cannot just add together the numbers of degrees, minutes, and seconds and write $12°75'98''$. Instead, we must take into account that $60' = 1°$ and $60'' = 1'$ to obtain the correctly represented answer of $13°16'38''$. Moreover, with more precise measurements, we might need to introduce halves of seconds, then tenths of seconds, and maybe even $71/532$ of a second; the numbers we need to use could become more and more complicated, to the point of being almost incomprehensible. A new method for representing these measurements was needed, and such a method was developed by the Dutch mathematician Simon Stevin (1548–1620) in the sixteenth century.

In 1585 Stevin published a remarkable book, *The Arithmetic of Simon Stevin of Bruges*, along with a twenty-nine-page appendix "De Thiende" (translated into English as "Disme: The Art of Tenths"). In *Arithmetic* Stevin made two important contributions. First of all, Stevin freely worked with negative numbers and realized that subtracting a positive number is the same as adding a negative number. In other words, 70 minus 40 is the same as 70 plus negative 40, that is, $70 - 40 = 70 + (-40)$.

Stevin's second contribution was to extend the concept of number beyond the positive counting numbers and their ratios. Stevin took quantity, or magnitude, as a basis for numbers. For Stevin a number was "the quantity of each thing," which means that a number is any magnitude associated with a quantity, for example a geometric length. Just as geometric magnitudes can vary continuously, so can positive numbers. Since, for Stevin, the negative numbers were simply the numbers that when added to another number give the same result as subtraction, the collection of all of Stevin's numbers was, in effect, what we call our number line.

Stevin was aware of the prevalent Pythagorean belief that unity generates number, and thus all numbers are produced from "one," and he was not insensitive to possible philosophical objections to his conception of number.[11] So Stevin adopted the Pythagorean perspective that all numbers are generated from some initial quantity, but replaced the discrete nature of generation of all numbers from unity by the continuous, geometric generation of all numbers from *naught*—conceived to be a point that he represented by the symbol 0. This one bold step reunited numbers with geometric magnitudes and removed mysticism from the answer to the question "what is a number?" With this shift in point of view, the formerly troublesome π and the square root of two could be accepted as numbers.

In the appendix "De Thiende" Stevin made the breakthrough that was needed for any practical use of the numbers from his continuum; he showed that each number could be represented in terms of powers of ten (and powers of one-tenth) using only the digits 0, 1, 2, 3, 4, 5, 6, 7, 8, and 9. Again, using our more codified notation, Stevin showed that every number can be written as a decimal, for example, $\frac{1}{2}$ = .5 and $6\frac{3}{4}$ = 6.75. The advantages of this system are readily apparent whenever a calculation is performed, just compare the ease of adding the three fractions $\frac{1}{2}$ + $\frac{3}{5}$ + $\frac{1}{20}$ or the equivalent decimals .5 + .6 + .05. Perhaps more importantly for us, Stevin showed that decimals could be used to approximate irrational quantities as closely as desired. For example, $\sqrt{2}$ is approximately 1.4, but a better approximation is 1.4142 and an even better approximation is 1.41421356. This process will never terminate because irrational numbers cannot be represented in Stevin's decimal system as finite expressions such as .234 or .99999.

Thus we are led to a second definition of a number:

A (positive) number is the length of a geometric magnitude.

SQUARING THE CIRCLE

The Greeks felt they better understood a geometric magnitude if it could be represented as the area of a square, so they developed methods for transforming different geometric figures into squares. But the Greeks only allowed certain methods for transforming one area into another. In our language these transformations had to be accomplished using only a compass, for drawing circles, and a straightedge, for drawing line segments. (The Greeks did not allow for the use of a ruler; remember that measurement was suspect.) Using these two tools, mathematicians showed that it is possible to transform any triangle into a square, and that it is possible to take two or more squares and transform them into a single square. Combining these ideas they could transform the area of any polygon into a square. (Any polygon can be decomposed into triangles; recall Plato's geometric chemistry. Each of these triangles can be transformed into a square, and these squares can be combined into a single square.) This idea offered the Greeks another way to understand π: Start with a circle whose radius equals one unit, so whose area equals π, and transform it into a square.

Alas, neither the Greeks nor anyone following them was able to discover the geometric construction that permits "squaring the circle." Partly because of the problem's apparent simplicity and partly because of the mystical connotations of the circle both professional and amateur mathematicians almost continuously studied this problem. The inability of anyone to accomplish this construction shrouded it in mystery. E. W. Hobson summarized these views in his book *Squaring the Circle: A History of the Problem* (1913): "The man of mystical tendencies has been attracted to the problem by a vague idea that its solution would, in some dimly discerned manner, prove as a key to a knowledge of the inner connections of things far beyond those with which the problem is immediately connected."[12]

Appeals to the failure of anyone to square the circle have made their way into literature. The earliest reference the author knows of was in the poetry of John Donne. In his poem "Upon the Translation of the Psalms by Sir Philip Sidney and the Countess of Pembrooke, His Sister" (1631) Donne wrote:

> Eternal God—for whom who ever dare
> Seek new expressions, do the circle square,
> And thrust into straight corners of poor wit
> Thee, who art cornerless and infinite.[13]

According to the scholar Roberto Bertuol, Donne's reference was a criticism of "man's attempt to put God, who is 'cornerless and infinite,' into 'strait corners of poore wit': that is, to understand God by means of our rational faculties."[14]

Later in the seventeenth century, the admittedly minor poet Margaret Cavendish (1623–73) appealed to squaring the circle to condemn, also in Bertuol's words, "men's striving to rationalize unknown elements such as nature's secrets and fancy by means of mathematics."[15] Cavendish began her poem "The Circle of the Brain Cannot Be Squared" (1653) with

> A Circle round divided in four parts
> Hath been great Study 'mongst Men of Arts;
> Since Archimed's or Euclid's time, each Brain
> Hath on a Line been stretch'd, yet all in vain.[16]

While Donne's appeal was to man's futility in rationalizing God, Cavendish's appeal was to man's futility in undertaking to mathematize nature and mind.

The mysterious relationship between a square and a circle continues to be employed by writers, in order to express both hidden truths and madness. In the short story "Goddess," Albert Wachtel invoked the dogged examination of this relationship to signal that one character, the mother in the story, is unable to cope with the society into which she has been thrust. When confronted with new knowledge, or a novel situation, that challenges what she believes or understands, the woman walks in ritualistic concentric circles and squares. The center of these combined geometric figures provides the woman with both physical safety and mental solace.

The reason neither the mother in Wachtel's story nor anyone else was able to square the circle is not because the square and circle have some incompatible mystical properties, it is because of the differences between the algebraic and geometric conceptions of number. In 1882 the German mathematician Ferdinand von Lindemann (1852–1939) established that neither π, nor the square root of π, is an algebraic number and thereby demonstrated the impossibility of squaring the circle.

The connection between algebraic numbers and geometric constructions is fairly simple. Using Descartes' coordinate system, squaring the circle means to take the circle centered at the origin whose radius equals

1 and produce a line segment whose length is $\sqrt{\pi}$, for example the line segment from (0,0) to $(\sqrt{\pi}, 0)$. However, if we begin with the circle whose radius is the line segment from (0,0) to (1, 0), using equations for lines and circles it is possible to show that any of the allowed, Greek, geometric constructions will lead to geometric magnitudes whose lengths are numbers involving combinations of square roots of rational numbers, and combinations of square roots of square roots of rational numbers, and so forth. All of these numbers are numbers in the algebraic sense. When Lindemann proved that the square root of π does not equal any algebraic number, he showed that no allowable geometric construction will ever permit you to begin with a line segment one unit long and construct a square whose area equals π (i.e., a square whose sides equal the square root of π). But even more importantly for us and our attempt to answer the question "what is a number?" Lindemann's result implies that the algebraic and geometric concepts of number do not coincide. Pi is a number in the geometric sense, but not in the algebraic sense.

THE MODERN CONCEPT OF NUMBER: PART ONE

This section's title is intended to indicate that it will not provide a definitive answer to the question "what is a number?" This is because a complete answer to that question depends on the modern understanding of mathematical infinity, which is the central topic of chapter 11. However, in this short section it is possible to provide an answer that will satisfy most working mathematicians.

So far we have offered two definitions for a number that overlap, but do not coincide:

> A number is a solution to an algebraic equation.

And

> A number is the length of a geometric magnitude.

Each of these definitions includes the counting numbers, fractions, some irrational numbers such as $\sqrt{2}$ and, if we accept the negatives of geometric magnitudes, the negatives of each of these values, but there are algebraic numbers that have no geometric interpretations, such as $\sqrt{-1}$, and geometric lengths that cannot be obtained algebraically, such as π. This possibly confusing situation is clarified if we take a slightly different point of view, the one almost universally taught in our schools, and begin with the *number line*, which is represented as a horizontal line with zero marked in the center—the positive numbers are to the right of zero and negative numbers to the left of zero. In the next chapter we examine the evolution of the relationship between a line and the points it contains, but here we take the naive point of view that each point on the number line corresponds to a number, and each of these corresponds to the length of a geometric magnitude—the line segment between zero and the point (or its negative, if the point is to the left of zero on the number line).

Although many of these points also correspond to solutions of algebraic equations, the *algebraic number* $\sqrt{-1}$ cannot be located on the number line because it cannot be interpreted as a distance. One way to introduce this quantity into the picture is to extend the number line to a *number plane* by taking another axis perpendicular to the traditional number line; this new axis is taken to represent multiples of $\sqrt{-1}$:

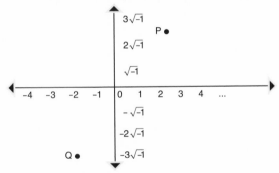

FIGURE 9.7. In the (complex) number plane each point can be completely described by its two coordinates. Here $P = (2.5, 2.5\sqrt{-1})$ and $Q = (-2, -3\sqrt{-1})$.

Early in the nineteenth century Gauss established an amazing fact about the number plane. To appreciate Gauss's result, we need to understand how to interpret points in this plane; the points P and Q, above, can be described by analogues of x-coordinates and y-coordinates. Yet there is another way to imagine these points—not as given by coordinates but as resulting from additions: $P = 2.5 + 2.5\sqrt{-1}$ and $Q = -2 - 3\sqrt{-1}$. In this interpretation, it can be shown that each of the points, P and Q, corresponds to an algebraic number: P is a solution to the equation $2X^2 - 10X + 25 = 0$ and Q is a solution to the equation $X^2 + 4X + 13$

$= 0$. However, not all points in the plane correspond to solutions to algebraic equations, for example the number $\pi + \sqrt{-1}$. Gauss' great theorem, which is so important that it is called *the fundamental theorem of algebra*, is

every algebraic number corresponds to
a point in the number plane.

In other words, the solutions to any polynomial equation, such as $3X^5 - 9X^3 + 1 = 0$, can all be written in the form $a + b\sqrt{-1}$, where a and b are numbers from Stevin's number line. Even more is known: the numbers a and b will themselves be algebraic numbers. Thus the fundamental theorem of algebra says that if we introduce into our number system only one algebraic number that does not correspond to a geometric magnitude, $\sqrt{-1}$, which is a solution to $X^2 = -1$, we can find the solutions to any algebraic equations by combining this number with algebraic numbers that are geometric magnitudes. Our possible discomfort with solutions to polynomial equations involving $\sqrt{-1}$ is somewhat relieved by their ultimate reliance on geometric magnitudes.

10 The Dual Nature of Points and Lines

As for what I have done as a poet ... I take no pride in it whatever. ... But that in my century I am the only person who knows the truth in the difficult science of colours—of that, I say, I am not a little proud.
— *Goethe*, Conversations of Goethe with Eckermann and Soret *(1875)*

When Stevin based his concept of number on that of a geometric magnitude, he relied on his understanding of a mathematical continuum to provide an intuition for number. Stevin agreed with Aristotle and took the continuum to be infinitely divisible; so, just as any geometric length can be divided into two others, between any two numbers there will be a third. All that is needed for a number to correspond to a geometric magnitude is for a line to contain points. The simplicity of this point of view glosses over the relationship between a line segment and points, and examinations of this relationship were important for both the invention of calculus and the establishment of the modern theory of mathematical infinity.

Georges Seurat's painting *Circus Sideshow* (Plate 2.2) was used to illustrate the purported appearance of the beautifully proportioned golden rectangle, but the most striking feature of Seurat's painting is not its geometric foundation. The first thing anyone notices is Seurat's painting style, known as the pointillist style, wherein paint is applied to the canvas not in strokes but in small dots of primary colors. While viewing *Circus Sideshow*, or almost any of Seurat's other paintings, we might imagine that Seurat was saying something about the fundamental nature of reality or our perception of it; however Seurat was not attempting to make a philosophical statement. Alternatively, we might guess that Seurat was exploring the wave/particle duality of light, having come down on the side of those who maintained that a ray of light consists of a stream of particles, but this position had fallen out of favor in the nineteenth century. Instead, Seurat's technique was more practical; his goal was to produce especially vivid colors, colors he did not believe he could obtain through mixing primary colors on his palette and then applying them to the canvas. Seurat was not exploiting the dual nature of a ray of light but the dual nature of color. These dualities have analogues in mathematics—there are two dualities concerning the nature of geometric magnitudes and another concerning the nature of points—but before we examine these, we briefly sketch two theories of light and color.

In the seventeenth century Isaac Newton (1642–1727) held a glass prism up to a window and observed the effect this filtering had on light. With his experiment, Newton discovered that the pure, white light entering the prism emerged as a spectrum of colored light, and the colors were always arrayed in the spectrum in the same order, from just barely visible red to orange, followed by yellow, green, blue, indigo, and violet. In order to confirm what he thought this meant, Newton then took two prisms and let the spectrum of colors emerging from one prism strike the other prism. What happened was that pure, white light emanated from the second prism: The first prism broke white light down into its constituent colors and the second prism recombined these colors to form white light. Indeed, that is exactly what Newton concluded— white light is a mixture of colored light.

There was a competing theory of color, going back at least to Aristotle, in which colors were thought to consist of combinations of light and dark. In this theory, which was developed by Johann Goethe (1749–1832) in his *History of the Theory of Colors* (1810), white light is more basic than colored light. Colored light is obtained by adding dark tones to the white light. One of Goethe's most influential conclusions was that our perception of a particular color depends on its context. Two adjacent colors interact—an effect called *simultaneous contrast*. The farther the two colors are apart on the color wheel, the greater the effect they have on each other. Newton probably would have attributed this phenomenon to some peculiarity of human physiology, but Goethe attributed this to the interaction of the colors themselves. More precisely, Goethe attributed simultaneous contrast to the interaction of the light and dark in the particular colors.

Goethe's theory was further elaborated upon early in the nineteenth century by the French chemist Michel-Eugène Chevreul (1786–1889). Chevreul was asked to investigate why certain dyes used in coloring carpeting and tapestries were not yielding colors as vivid as was hoped for. He discovered that some of the colors in the textiles were dull not because of the particular dyes being used, but because of the juxtaposition of colors. This insight led Chevreul to formulate a theory of color interactions that was much more subtle then Goethe's. For example, Chevreul discovered that adjacent dots of blue and yellow are perceived as a more vivid green than a splash of green paint that has been obtained by mixing blue and yellow paints on a palette. In effect, the eye, or more precisely the brain, averages the primary colors and perceives the intermediate ones. Chevreul published his results in *The Laws of Contrast of Colour* in 1839.[1]

One of the first artists to acknowledge Chevreul's influence on his work was the British painter Joseph Mallord William Turner (1775–1851). Turner is known for his use of contrasts, such as in his *Shade and Darkness—The Evening before the Deluge* (1843), and some of these contrasts are obtained by applying small splashes of primary colors onto his larger brush strokes. Half a century later, Seurat did not use any brush strokes; he employed only small dots of pure color.

The dual nature of light is more widely known than the dual nature of color. Newton had proposed that a ray of light is made up of a stream of corpuscles, but there was the competing view, espoused by Newton's contemporary Robert Hooke (1635–1703), that a ray of light is a wave. Then, in 1801, Thomas Young (1773–1829) rediscovered something Newton had overlooked—light passing through a narrow slit produces a wavelike interference pattern. (Francisco Grimaldi [1618–63] had discovered this in the seventeenth century but his work was not widely known.) Young's experiment showed that a ray of light does not behave like a stream of particles but like a wave oscillating through space—or the proposed ether. The mathematics behind the wave theory of light was fully developed late in the nineteenth century by James Maxwell (1831–79), which could have settled the matter. Whatever the nature of light, it was assumed that Maxwell's equations would provide a sufficiently predictive model. However, just a few years later, early in the twentieth century, Einstein revitalized the particle theory of light with his explanation of the photoelectric effect. With Einstein's discovery came the understanding that light can, simultaneously, behave like a stream of discrete particles and like a continuous wave. Which of these qualities is detected depends on which aspect of light is being considered.

Whether light should be thought of as a stream of particles or as a vibrating wave, and whether colored light should be thought of as a constituent of white light or a mixture of white and black light, depends on what is being sought. Newton's particle theory did not help physicists understand the propagation of light, and Goethe's theory of color did not provide a basis for quantifying the differences between red and violet. But the particle theory does explain the interaction of light with matter, and simultaneous contrast does offer a basis for a theory of color that can be exploited by artists.

NAIVE MATHEMATICAL GEOMETRY

The most basic of all geometric objects are line segments and points, and given their central roles in geometry, it would seem that they would be the best understood of all geometric entities. Surprisingly, they are not now nor have they ever been. Both a line segment and a point possess dual natures, which are analogous to both the dual nature of light and the dual nature of color. The best way to begin to understand the relationship is to return to Euclid's *Elements*.

Euclid's definition, or perhaps description, of a point is essentially that of an indivisible atom:

A point is that which has no part.

Whether this statement offers a definition or a description of a point, it does capture the idea that a point is the smallest imaginable mathematical entity. Euclid defined a line segment in two steps. Euclid first defined a line, which can be either straight or curved:

A line is a breadthless length.

Euclid then explained what it means for a "line" to be "straight":

A straight line is a line that lies evenly with the points on itself.

Euclid did not claim that points comprise a line; the concept of a geometric length does not depend on that of a point. A "length" is an undefined concept. All that Euclid acknowledges is that a line has "points on itself."

This leads naturally to questions that appear to be mathematical analogues of

How many angels can dance on the head of a pin?

But these questions are not frivolous, they are relevant to understanding the evolution of mathematical thought, and during the Middle Ages and Renaissance, these questions were thought to be relevant to understanding the material world.

If we postpone for now questions about the nature of points or lines, the next question is

Does a line segment simply contain
points or is it made up of points?

The follow-up question, regardless of how the first one is answered, is

How many points are in a line segment?

In order to overcome both Pythagoras' equivalence of the mathematical and material worlds and Plato's otherworldliness, Aristotle took mathematical objects to be abstractions from the physical world. And because they are abstractions from the physical world, mathematical objects were thought to be relevant to our understanding of the physical world. Aristotle, and anyone adopting his physics, believed that there was a correspondence between physical magnitudes and geometric magnitudes. So, whatever the nature of the continuum in one realm, it would be the same in the other.

Aristotle answered both of the above questions. His answer to the second was that the collection of points contained in a line segment was potentially infinite by division. To answer the first question, Aristotle sought to clarify the relationship between the points contained in a line segment and the line segment itself. Aristotle thought of a line segment as a continuum having an existence beyond the points it contained, and in his *Physics*, Aristotle gave an argument intended to show that a continuum cannot be made up of small objects having no parts. His argument is known as the *contact argument*: If a continuum is made up of indivisible elements, then these elements must be in contact with one another. They cannot be joined together side by side so as to produce, say, a line, because if they have sides then they have parts and are not indivisible. So, Aristotle continued, these indivisibles must be wholly in contact with one another, and if two indivisibles are joined whole to whole they occupy the same space. Thus no number of indivisibles can be put together to produce anything but an accumulation of indivisibles in the same place.[2]

Transferred over into the physical world, Aristotle's contact argument does not refute the possibility that there could exist physically indivisible atoms that are somehow bound together to produce material substance. In the Leucippus/Democritus atomic theory, atoms were physically indivisible pieces of matter. They were not conceptually indivisible in that they were units without parts; they simply had the material property of being indivisible. The contact argument argues against the possibility that conceptually indivisible entities, like Euclid's points, could hold together to form anything. Physical atoms were rejected by Aristotle's physics, not by his understanding of mathematical lines, and since Aristotle's natural philosophy was the dominant one, almost everyone rejected the existence of atoms.

THE DUAL NATURE OF LINE SEGMENTS

Neither Aristotle's answers to the two questions above nor his conception of a point went unexamined. By adopting views contrary to Aristotle's, mathematicians of the Renaissance and Enlightenment made very significant progress in understanding geometric entities that were not available to the Greeks, such as areas enclosed by fairly exotic curves. Mathematicians embraced these views not for their philosophical purity, but for their utility. Before the Renaissance, long before any significant mathematical progress had been made in Europe, the nature of geometric magnitudes was examined for mostly theological reasons. We examine some of these fourteenth-century debates in the next section, but in order to appreciate the difficulties theologians, and then mathematicians, faced, let's first examine one duality in the modern conception of a line segment.

The modern point of view is that points make up a line—in fact, a line is defined as a collection of points satisfying a certain property. (There is a circularity in this reasoning because the property that the points satisfy is that they make up a line segment; thus a line segment is ultimately an undefined concept.) And although a line is made up of points, it does have an existence beyond those points; for example, it has a length, which is something that could not be predicted from the individual points. However, each point in a line somehow knows it is contained within a line; it knows how to behave in order that the aggregate of all points holds together to form a line. This strange property of the points contained in a line segment has an analogy in the relationship between photons and the light ray that contains them.

In *The Elegant Universe*, Brian Greene provides a very clear explanation of how we can demonstrate that a ray of light has an existence beyond the photons it contains. The experiment, which is an elaboration of the one Young used in 1801 to show that light behaves like a wave, shows that photons in a beam of light somehow "know" that they have not only an individual but also a collective existence. The basic outline of this experiment is simple—it involves shining light through a metal plate having two vertical slits cut through it. If the ray of light consisted of a beam of particles, then we should observe the pattern on the left in Figure 10.1.

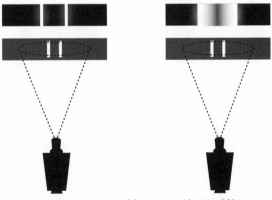

FIGURE 10.1. Two possible patterns that could be produced by light shone through two parallel slits.

But this is not what the experiment showed. When Young shone light through a plate with two vertical slits he saw the pattern on the right. This pattern is the wave interference pattern that is obtained, for example, when two stones are simultaneously dropped into a pond; what the pattern indicates is that the ray of light passing through the slit on the left, and the ray of light passing through the slit on the right, interfere with each other. The two rays of light act like waves, rather than like what we imagine to be straight lines.

In the next stage of the experiment, the ray of light was focused on the plate with two parallel slits, but the light is sent to the plate one photon at a time. Since there are no two photons being sent simultaneously, they should not interfere with each other; so, after sending a large number of these photons we would expect a pattern such as the first one above. But this is not what happens. Each photon knows it is part of a greater entity and so each behaves in the way it should, given its lot in life. The pattern that emerges is the second one; this pattern shows that the photons act like they are being interfered with, even though they are not.

The mathematical analogue of this experiment can only be imagined. In this experiment we would start with a line segment, lift points from the line segment one at a time, and throw them onto a piece of paper. The points would not form a random pattern on the paper, but would instead line themselves up. The more points we take from the original line segment, and throw at the paper, the more fully the line segment will have been transferred.

This first duality associated with line segments concerned the nature of the continuum and is analogous to the wave/particle duality of light rays. But just as there is a second duality associated with color, there is a second duality associated with points—points can be thought of as dimensionless dots or as indivisible elements (with or without meaningful size). Each of these conceptions of a point has a place in modern mathematical thought; before we discuss this we return to the fourteenth century.

THE FOURTEENTH-CENTURY DEBATES

Several reasons have been offered for the fourteenth-century reemergence of atoms. The least intriguing of these is the assertion that the syllogistically oriented philosophers and Scholastics of the late Middle Ages were simply looking for loopholes, or logical flaws, in Aristotle's reasoning; the antagonism between philosophy and theology continued well past the Condemnation of 1277. This view underestimates the role of religion as a motivation in these discussions and implies that the Scholastics had no emotional connection to their conclusions. More attractive explanations for the return of atomism associate it with the Scholastic attempts to reconcile Aristotelian principles with theology, leading to arguments for the consistency of ideas contrary to Aristotle's with those of his that were widely accepted.

One difficulty with Aristotelian principles that proponents of the existence of atoms of indivisible physical elements sought to overcome concerned the nature of angels. Angels were believed to consist of an incorruptible substance. Implicit in the purity of angelic substance was the belief that this substance could not be a combination of the still-accepted fundamental terrestrial elements: earth, air, fire, or water. More generally, the incorruptibility of angelic substance meant that it could not be made up of more basic materials or parts. Thus, whatever the imagined state of the material existence of angels, they were not only physically indivisible entities, but could not even be imagined to be divisible. Thus angels were seen to be conceptually indivisible.

Alas, this understanding of angels presented theologians with a quandary. In his analysis of the nature of motion, Aristotle had demonstrated that if space and time were infinitely divisible, then a conceptually indivisible entity would be unable to move. Aristotle reasoned as follows: Suppose, for the sake of argument, that indivisibles exist, and imagine that there is an indivisible located at a position ab that then moves to position bc. (In Figure 10.2 the indivisible is represented by the triangle.)

FIGURE 10.2. The movement of an indivisible element from position ab to position bc.

Aristotle then posed the question: Where is the indivisible during the time interval between when it was at ab and when it is at bc? Aristotle's answer was that the indivisible must be in some intermediate position, wherein it is partly at ab and partly at bc. This means the indivisible has distinguishable parts, the part that at the intermediate time is at ab and the part that at the intermediate time is at bc, and so the indivisible is not a unity. It is supposed to follow that a conceptually indivisible entity cannot move.[3]

The atomist response to Aristotle's simple argument was not to refute it but to point out a presumed fallacy in Aristotle's reasoning. It was almost universally held that time was infinitely divisible, the belief that there might exist indivisible pieces, or units, of time having been espoused only by some Epicureans, followers of the third-century B.C. philosopher Epicurus, and by modern physicists. Assuming the infinite divisibility of time, Aristotle's argument could not be dismissed on the grounds that there were not some intermediate moments between when the indivisible was at ab and when it was at bc. What was wrong with Aristotle's reasoning, according to the atomists, was that there is no intermediate position between position ab and position bc. They simply asserted that space is not infinitely divisible, so at any moment the indivisible was either at ab or at bc—the triangle instantaneously switches from position ab to position bc.

Another theological argument for the existence of indivisibles relied on God's omniscience: If a continuum, or chunk of matter, were to be infinitely divisible, then even God could not know how many parts it had. Augustine's contention in *The City of God* that God could see an infinitude as a completed whole did not address the problem of how many parts infinite divisibility would produce.[4] This conundrum elicited various responses. It was argued that the infinite divisibility of matter did not mean that at some time it could be infinitely divided. It was also argued that the infinite divisibility of matter meant that it did not have parts, and so there were not any objects for God to count. But the simplest way to address this was to conclude that this was not an issue because matter simply is not infinitely divisible; matter consists of finitely many indivisible parts. These two theological arguments lead to the conclusion that a physical continuum, whether it is a chunk of space or a piece of matter, is made up of a finite number of indivisible elements.

Although the atomists may have been motivated by theological considerations, they often based their arguments for the existence of atoms on mathematical or, less often, philosophical principles. What is especially interesting is that all of these discussions were intended to either defend or refute the existence of conceptually indivisible units, as opposed to the physically indivisible atoms of the Leucippus/Democritus atomic theory. In the fourteenth century a material atom was infinitely small, without size or mass, and so more like a mathematical point than a piece of matter. As the scholar Laurence Eldredge asserted in his discussion of the meaning of "point" in Middle English literature, whatever principles were used to frame the argument, the atomists were not as convincing as the infinitists.[5]

A MATHEMATICAL REFUTATION OF INDIVISIBLES

The claim that a physical continuum can be made up of finitely many indivisibles was fairly easily challenged—it was challenged by an appeal to the same idea that Duns Scotus used to prove that time cannot possibly be cyclic—the incommensurability of the side and diagonal of a square. Without acknowledging the generally accepted correspondence between material and mathematical continua, in the thirteenth century Roger Bacon (1214–92) sought to undermine the claim that a chunk of matter is composed of finitely many indivisibles by showing that a finite line segment cannot be made up of finitely many indivisible mathematical entities—mathematical atoms: "If lines are composed of atoms, the diagonal of a square and its side will have the same ratio as the number of whole atoms making up these lengths; therefore these lengths are commensurable, contrary to what the mathematicians teach."[6] The details of Bacon's argument are simple: Let D denote the length of the diagonal of the square, let L denote the length of a side of the square, and assume that each indivisible element has size ε. If the side of the square contains M points and the diagonal of the square consists of N points, then the length of the diagonal is $N \times \varepsilon$, and the length of the side is $M \times \varepsilon$. Thus $D = N \times \varepsilon$, and $L = M \times \varepsilon$, so

$$D/L = (N \times \varepsilon)/(M \times \varepsilon) = N/M.$$

Rewriting the equation, above, without the middle term we obtain

$$D/L = N/M.$$

Through this argument, Bacon has shown that if every line segment is made up of finitely many indivisibles, then the diagonal and side of a square are commensurable. But this is precisely what the Pythagoreans had shown to be false. (This can be rephrased numerically. Suppose

Bacon's square had sides of length 1; if segments consist of points, then $\sqrt{2}/1 = M/N$. Thus $\sqrt{2}$ is a rational number, contrary to the Pythagorean discovery.)

The preceding incommensurability argument shows that a line segment cannot be composed of a finite number of indivisibles. However, as there cannot be an existing infinite collection, a line segment, or a chunk of space or matter, cannot consist of an infinite number of indivisibles. Thus we are left with Aristotle's infinitely divisible conception of a continuum.

Atomists attempted to refute each of the arguments: That the number of indivisibles that make up a line segment is related to its length and that there cannot exist an infinite collection. Walter Chatton (c. 1290–1343) addressed the first claim and Henry of Harclay (c. 1270–1317) the second.

CHATTON'S GEOMETRIC WAVES

Chatton deserves recognition for his creativity in countering geometric arguments such as Bacon's. Chatton was an atomist, and Chatton's adoption of the existence of indivisibles emerged from his desire to avoid the possible existence of a greater and lesser infinity. Chatton imagined that if two unequal continuous magnitudes were infinitely divisible, then they must contain the same number of parts, which, for Chatton, implied that the magnitudes must necessarily be the same size.

First, let's review the argument Chatton had to overcome, which purports to show that if circles are made up of finitely many indivisibles, then all circles must contain the same number of indivisibles and so must have the same size. The entire argument is illustrated in Figure 10.3.

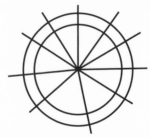

FIGURE 10.3. Radial lines between two concentric circles that show there is a correspondence between the points on the smaller circle and the points on the larger circle.

The radial lines are drawn as follows: Draw a line from the common center of the two circles through each point on the smaller circle. Each of these lines will intersect the larger circle at a point. Once this operation has been completed, there will be a line connecting every indivisible on the smaller circle with an indivisible on the larger circle. This process cannot miss any of the indivisibles on the larger circle; if it did, then a line drawn from that point to the common center of the two circles would intersect the smaller circle in a point that has not been accounted for. Thus, so the argument goes, these two circles contain the same number of indivisibles and so must be of the same size.

We saw in chapter 4 that it is possible to arrive at conflicting conclusions by appealing to the same evidence if you start with different assumptions. In order to prove that the sun is close to the earth, Cosmas assumed that the earth is flat and appealed to exactly the same evidence, the stick's shadow, that convinced Eratosthenes that the earth is not flat. The mathematical argument above purported to show that if the circles consist of finitely many indivisibles, then they would have to be the same size. Chatton reversed this argument; he used the two concentric circles to prove that if the circles do not consist of finitely many indivisibles, then they must be the same size.

Chatton first argued that if the circles consisted of infinitely many indivisibles, then they would have to be the same size, else we would have two sizes of infinity. Chatton then argued that if the circles were infinitely divisible, then, as there is only one size of infinity, the two circles would contain the same number of divisions, so points, and so the two circles must be the same size. In making his argument, Chatton appealed to the process of division and maintained that if, in the process of dividing the circles, there is a correspondence between the divisions of one circle and the divisions of the other, then the two circles must, ultimately, have the same number of divisions. In other words, we must be able to extrapolate from the correspondence between the divisions of the two circles, which emerges in the process of subdividing them, to a conclusion about what happens when the process of division has been completed. Thus, Chatton assumed that the process of subdividing the circles could be completed, so at some time all of the parts will independently exist.

Once Chatton had determined that a geometric continuum consists of a finite number of points, he needed to undermine Bacon's mathematically sound argument. This is where Chatton showed his creativity: Chatton claimed that geometric shapes are not the two-dimensional objects we imagine them to be but are three-dimensional and occupy space. In this way, geometric objects have attributes, or parts, beyond what we had believed; for example, a surface has not only length and width but also a front and a back. So a point on a line might appear to be in the same plane as the line but could lie behind the line or in front of it. Chatton concluded that any drawing seeming to show the side and diagonal of a square to be commensurable, or even to contain the same number of points, is an illusion. (One way to think about this is to imagine that the diagonal is a wavy line, with peaks and troughs above and below the plane of the drawing.) This means that we can never know how many indivisibles are on a line segment, so we cannot conclude anything about the ratio of the number of indivisibles on the side and diagonal of a square.

HARCLAY'S INFINITE COLLECTIONS

Not all proponents of atomic theories were so willing to offer innovative mathematical or theological arguments. A good example of a fourteenth-century philosophical argument for the existence of atoms is the one given by Henry of Harclay, who became chancellor of Oxford University in 1312. Atoms were necessary to Harclay's philosophical understanding of existence. In his view, anything that exists in the material or mathematical world must be a singular entity; something that cannot exist independent of being part of something else does not have the property of existence itself. In the process of subdividing a continuum, one must obtain parts, and if these parts can be subjected to further division then these parts have existence beyond being a piece of the larger object. From this, Harclay concluded that if matter is infinitely divisible, one must obtain smaller and smaller pieces whose only existence comes from their being parts of a larger whole, contrary to the very nature of existence.

Harclay understood that finitely many points could not make up a segment, so he was led to accept the conclusion that a continuum must contain infinitely many points. Then, as two line segments can be of different lengths, if each contains infinitely many points, there must be different sizes of infinity. Harclay simply accepted this conclusion. However, Harclay had to address Aristotle's contact argument, which showed that indivisibles could not make up a continuum. Harclay dismissed Aristotle's argument with the claim that atoms do not contact each other whole to whole, and so never amount to anything larger than a single indivisible, but reside near one another, mysteriously, "in respect to distinct locations."[7] This was perhaps every bit as convincing as Aristotle's original argument that atoms cannot make a whole because if they are in contact they are either coincidental or have parts. And Harclay gave a theological reason for why a line segment must consist of points sitting side by side: "It is certain that God knows every point that can be designated in a continuum. Take, then, the first inchoative point on a line. God perceives that point and any point in this line different from it.... It follows, then, that either up to that more immediate point which God sees there intervenes some line ... or one does not."[8]

In other words, there cannot be a line segment connecting the first point God perceives on the line segment and the next point on the line segment that God perceives. The first point God sees and the second point God sees are adjacent. (It is important to note that Harclay's conception of a geometric continuum is contrary to one of the most basic tenets of Euclidean geometry, that two points always determine a line. The Euclidean axiom is not that two nonadjacent points determine a line, but that any two points determine a line.)

MATHEMATICS AND INDIVISIBLES

None of these arguments about the nature of either matter or line segments were definitive. One early fourteenth-century infinitist offered perhaps the strongest rejection of the atomists' indivisibles by invoking the law of sufficient reason in a form that has come to be known as *Occam's razor*. William of Ockham (c. 1288–c. 1348) refuted all of the atomists' arguments with a single observation: There is no need to assume that a continuum is made up of points because everything in mathematics and physics works out perfectly well without that assumption.

But Harclay's conception of a line segment—that it consists of a string of infinitely many points, like beads on a taut line—is reflected in the thinking of several important Renaissance mathematicians. Three of particular note are Galileo Galilei (1564–1642), Johannes Kepler (1571–1630), and Bonaventure Cavalieri (1598–1647). Galileo is, of course, better known for his conclusions about the physical world, but he was a professor of mathematics and also made contributions to mathematics. For our purposes his most important contribution was the point of view he passed on to his student Cavalieri.

Galileo, with Harclay, believed that a geometric continuum is made up of a string of infinitely many indivisible points. Galileo was, of course, a very pragmatic person; when he saw spots on the sun he simply accepted it as fact. Indeed, his entire conception of science was to seek to answer not *why* but *how*. As an application of his temperament in mathematical thinking, consider how Galileo is said to have resolved the paradox of the two circles. When asked why the two circles are not the same size, he is said to have responded that there must be larger gaps between the indivisibles in the larger circle than in the smaller one.

Cavalieri took on the project of applying the theory of indivisibles to solve geometric problems. In his *Geometry Advanced by a Thus Far Unknown Method, Indivisibles of Continua* (1635) Cavalieri proposed what has become known as "Cavalieri's principle." Imagine starting with fourteen thin, rectangular strips of paper, all the same size, and arranging them into two groups of seven, as in Figure 10.4.

FIGURE 10.4. Cavalieri's conception of solid objects as being made up of infinitesimally thin rectangles.

If we view each of these groupings as a two-dimensional area, then they would be equal since they are composed of the same number of identical rectangles. Cavalieri's principle is the reversal of this process, where areas are imagined to consist of infinitely many horizontal line segments.

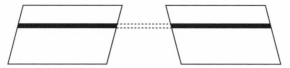

FIGURE 10.5. Cavalieri's principle says that if two areas
have equal cross sections then they are equal.

If the corresponding segments from the two areas were the same length, then the two areas would be equal. One consequence of this point of view was Cavalieri's conclusion that an area or volume would require an infinite (or indefinite) number of indivisible elements; for example, an area would consist of infinitely many parallel lines, but he did not explain what he meant by this.

Kepler took a slightly different view of geometric objects; he adopted a point of view that had been espoused by Nicolas de Cusa in the fifteenth century. According to de Cusa, and Kepler, a line segment, or a curve, is made up of infinitely many infinitesimal lengths (not points). An infinitesimal is not conceptually indivisible, like Cavalieri's indivisibles, but just *infinitely small*. Kepler illustrated the utility of his point of view by using infinitesimals to find the area of a circle: Since the arc of a circle is made up of infinitely many infinitesimal lengths, drawing radii from the center of the circle to the two ends of each infinitesimal reveals that the interior of the circle consists of infinitely many triangles. These triangles can be rearranged with their infinitesimally short bases along a straight-line segment; the length of this segment will be the circumference of the circle. Since the area of a triangle is determined by its base and height, the total area contained in the triangles arranged along the segment can be found, and it equals the area of the circle.

In the last third of the seventeenth century, both Newton and Gottfried Leibniz (1646–1716) combined extensions of the ideas of Cavalieri and Kepler to establish what is now known as calculus. We will only consider how the techniques of calculus permit calculations of areas bounded by curves, but its range of applicability is substantial, including relating acceleration, velocity, and distance traveled for a freely falling body, finding tangent lines to curves (needed to design lenses), maximizing and minimizing quantities (such as determining the optimal angle of inclination of an artillery cannon), and finding lengths of curves (for example how far a planet has traveled in a few days).

Both Newton and Leibniz based their theory of calculus, at least so far as the calculation of areas goes, on ideas that were formalized in the nineteenth and twentieth centuries. The nineteenth-century point of view is illustrated in Figure 10.6.

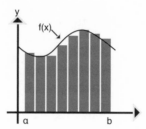

FIGURE 10.6. Approximating the area under a curve using rectangles.

The area under the curve, where the y-coordinate of each point on the curve is given by a formula $y = f(x)$, is approximated by the sum of the areas of the rectangles in Figure 10.6. Newton imagined that the base of each rectangle was small and then, relying on the infinite divisibility of the continuum, calculated what happened when x became zero. Leibniz, on the other hand, thought of the base of each rectangle as an indivisible, so the area under the curve was approximated by an infinite sum of these areas. To obtain the precise area under the curve, Leibniz worked almost algebraically with these infinitesimals. Astonishingly, especially to those who were critical of either approach, both of these calculations seemed to work! In the nineteenth century, mathematicians finally established the theoretical foundation for Newton's approach (see chapter 11); not until the middle of the twentieth century was Leibniz's method fully understood (see chapter 12).

A SECOND DEFINITION OF MATHEMATICAL INFINITUDE

Aristotle had defined an infinite collection as one that always has some part outside itself; a collection is infinite if no matter how much of it has been accounted for, it is not completely accounted for. What emerged from Gregory of Rimini's analysis of the arguments put forth by Burley was a new understanding of the quantitatively infinite that was remarkably close to the modern view (see chapter 5). Aristotle's actual/potential distinction was replaced with a subtler one. Before offering a simple explanation of this differentiation, we phrase it as the Scholastics did. In the mid-thirteenth century, Peter of Spain, who may have gone on to become Pope John XXI, provided the following definition: "Infinite is taken in two ways; in one way it is taken categorematically, significantly as a general term, and thus it signifies the quantity of the thing which is subject of predicate, as when one says, the world is infinite. . . . In another way it is taken syncategorematically, not insofar as it indicated the quantity of the thing which is subject or predicate, but insofar as the subject is related to the predicate, and in this way there is distribution of the subject and [it is] a distributive sign."[9]

The quotation is written in the impenetrable Scholastic style, perhaps accounting for the five-hundred-year gap between these preliminary discussions of the infinite and the nineteenth-century breakthrough, but its meaning can be clarified through an example. Consider the sentence, "This rocket travels infinitely fast." There are two ways in which this sentence can be understood and both depend on how the word *infinitely* is being used. The first meaning, wherein "infinitely" is being used *categorematically*, is that the rocket travels faster than anything that travels with a finite velocity. That is, no time passes while the rocket moves from one place to another. The second meaning is that given any velocity, be it one mile per hour, 1 billion miles per hour, or 1 billion billion miles per hour, the rocket can travel faster than the specified velocity. In this second interpretation, "infinitely" is being used *syncategorematically*. In the first interpretation, "infinitely" is understood as exceeding all finite magnitudes or quantities; in the second, as exceeding any given or specified magnitude or quantity.

This distinction between the syncategorematically and categorematically infinite was not available to Aristotle because a basic tenet of Aristotelian physics is that whatever can be said, or imagined, to exist potentially will exist actually, and that whatever cannot exist actually cannot exist potentially. This is precisely the principle Aristotle invoked to deny the existence of a vacuum. Applied in the present context, this principle becomes: If a body can potentially become greater in size than any given magnitude, and so be syncategorematically infinite, it would then be, in actuality, greater than any given magnitude, and so would be an existing infinitude and thus be categorematically infinite.

Gregory went on to reinforce the acceptance of this distinction by reformulating it: A categorematically infinite magnitude is that which is larger than any finite magnitude and a syncategorematically infinite multitude is "more considerable" than any finite magnitude. In his conception of the categorematically infinite, Gregory anticipated the discoveries mathematicians made more than five hundred years later: The categorematically infinite transcends finitude. The reason this is such an important idea is that it allowed Gregory to reject the powerfully intuitive notion that one infinity cannot be larger than another, and the important step was for Gregory to reconsider the very notion of a whole and a part. Gregory wrote that there are two ways to consider the terms whole and part. We restrict our attention to infinite multitudes although the same terms apply to infinite magnitudes: "According to the first way, any multitude is a whole with respect to another multitude when the first multitude contains the second . . . and when it contains, in addition, an object or objects distinct from all and each of these. . . . According to the second way, in order for a multitude to be *whole* with respect to another multitude, it has first to contain the second multitude, as in the first way; in addition it has to contain a determinate number of things of determined magnitudes . . . which are not contained in the contained magnitude." Gregory then applied these concepts to the Aristotelian principle that there cannot exist a greater and a lesser infinity: "In [the first] way, an infinite multitude can be a part of another infinite multitude. . . . In the second way, an infinite multitude cannot be either *whole* or *part* of another infinite multitude [because there cannot be a determinate number of things contained in the whole but not the part]."[10] Alas, Gregory never provided a clear meaning for an infinite multitude, and so his belief that one infinity could be contained in another infinity was not accepted and was not even understood until the nineteenth century.

11

Modern Mathematical Infinity

Self-evidence is often a mere will-o'-the-wisp, which is sure to lead us astray if we take it as our guide. For instance, nothing is plainer than that a whole always has more terms than a part, or that a number is increased by adding one to it. But these propositions are now known to be usually false.

— Bertrand Russell, "Recent Work on the Principles of Mathematics" (1901)

In the above quotation Russell refers to the purportedly self-evident truth—the whole is always greater than a part—that led to the paradoxes surrounding the infinite. But Russell also mentions another seemingly obvious truth and alludes to it not being universally correct, that when one is added to a number the result is a larger number. Indeed, by looking at examples, such as by adding one to the numbers 3, or $\frac{1}{4}$, or $\sqrt{2}$, it does seem to be the case that the result is always a larger number. So Russell must have something else in mind; he reveals what that is in the next sentence:

> Most numbers are infinite, and if a number is infinite you may add ones to it as long as you like without disturbing it in the least.[1]

To understand the mathematics behind Russell's claims, we will reexamine the meaning of the counting numbers. We will see that Russell's conception of number depends on the same idea as the modern interpretation of Stevin's decimal system and the fractals of modern chaos theory. The underlying principle is that it is possible to complete some processes that require an unlimited number of tasks. We begin with geometry.

GEOMETRIC MAGNITUDES (REVISITED)

Stevin's sixteenth-century identification of numbers with geometric magnitudes did not provide a system of numbers sufficient for solving all algebraic equations, but it did offer a number system that allowed for the comparison of the lengths of any two geometric magnitudes. Of course, even the ancient Greeks allowed for the comparison of any two geometric magnitudes, they just did not do this by attaching numerical quantities to the lengths. However, two things happened to undermine our certainty that we understood geometric magnitudes. The first was the discovery that the geometry of space is not the one handed down to us by the Greeks, which definitively separated the mathematical and material worlds. The second was the complementary realization that our misleading intuitions about physical space do not necessarily lead to a well-founded notion of a straight line or flat surface. Thus, Stevin's concept of number, which requires the unquestioned reliability of geometric truth, might suddenly seem less convincing.

Rather than devolve into another philosophical discussion, we tackle the simplest problem associated with defining numbers to be geometric magnitudes: Given a line segment connecting two points, how do we determine its length? Thanks to Descartes' introduction of coordinate geometry in the seventeenth century, this problem does have a simple solution in Euclidean geometry. "Given two points" means we are given the position of two points, and therefore their coordinates, from which it is easy to calculate the length of the line segment between them. But just because we can measure a line segment, does this mean that every geometric magnitude will have a well-defined length? For example, how does one calculate the length of a portion of a parabola?

It is basic to Stevin's concept of number that a portion of the parabola has a well-defined length; it just might not be easy to determine. (Recall how much work Archimedes had to do in order to estimate the circumference of a circle whose diameter is one unit, and so π.) In order to estimate the length of any curve, it has to be imagined as lying in Descartes' coordinate plane (so each point on the curve is described by an x-coordinate and a y-coordinate). Then the process of estimating the length of the curve is reminiscent of Archimedes' estimation of the circumference of a circle, and we illustrate this process below.

Archimedes estimated π by using polygons to approximate the circumference of a circle whose diameter is one unit, so whose circumference is π. Archimedes started with polygons having 6 equal sides and then successively doubled the number of the sides of the inscribed, and circumscribed, polygons to 12, 24, 48, and 96. Let's just think about the inscribed polygons. At each stage in the process of successively doubling the number of sides of the polygons, the length of the perimeter of each successive polygon is a better approximation of π, but the value will never precisely equal π because the polygon will never exactly be the circle. Even if Archimedes' procedure is continued, and the perimeters of inscribed polygons with 192, 384, and 768 sides are calculated, the value will never equal π.

The length of any curve that is not "infinitely long" can also be estimated by adapting Archimedes' method for estimating π. Suppose we have a curve in the plane that connects two points, for example, a curve that looks like a large S. We assume, keeping in mind Stevin's definition of number, that the curve has a well-defined length, which we denote by L. To obtain a first estimate of L we cannot use inscribed polygons, both because the way the curve twists and because the curve has no inside. Instead, we reinterpret Archimedes' inscribed polygons by thinking of each inscribed polygon not as a polygon but as a collection of line segments approximating the circle. And since we know how to calculate the length of a line segment, we can get an approximation of the *length* of the circle.

We use this idea of approximating line segments to calculate the length of a portion of a parabola; we will choose more and more, shorter and shorter line segments at each stage. Archimedes started with six line segments, the sides of a hexagon, but we begin more conservatively, with one line segment. The first thing we have to do is view the parabola in the Cartesian plane. The parabola is given by the equation $Y = X^2$; we will find the length, L, of the parabola that connects the points $(0, 0)$ and $(1, 1)$. This may be accomplished by first estimating the length by approximating the arc by the line segment connecting the points $(0, 0)$ and $(1, 1)$, as on the left below, and then by approximating the arc by two line segments, one connecting the points $(0, 0)$ and $(.5, .25)$, and another connecting the points $(.5, .25)$ and $(1, 1)$, as on the right below:

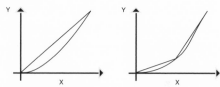

FIGURE 11.1. It is possible to approximate the length of the parabola $Y = X^2$ between the points $(0, 0)$ and $(1, 1)$ by using line segments. On the left, one segment is used, and on the right, two segments are used. The more segments used the better the approximation.

The first segment has length $\sqrt{2}$, which has an unending decimal representation that begins with the digits 1.41421. The sum of the lengths of the two segments on the right is a better approximation of the length of the arc; this sum is again an unending decimal but it starts with the digits 1.46040. Repeating this process one more time, and estimating the length of parabola using four line segments, yields the estimate 1.47428.

If we could (somehow) successively double the number of segments and produce these sequences of segments at a faster rate, and even more unlikely, compute the sum of their lengths at a faster and faster rate, we would in the end have calculated L. One of the great accomplishments of Newton and Leibniz is that they showed how this seemingly impossible task could be accomplished, at least for curves that are defined by fairly simple formulas. Instead of taking their approach, we want to consider how it could even be imagined to be possible to obtain these increasingly accurate approximations to what should be the length of L.

ZENO'S PARADOX (REVISITED)

In Zeno's thought experiment the Dichotomy, it is impossible for anything to move from one position to another, say from A to B, because the movement is incremental, and at each step the new increment is less than what would be needed to complete the journey. We examine this paradox one last time under two different assumptions concerning the way in which you move when going from A to B. We first assume that you move from halfway point to halfway point and take the same amount of time for each step, and second, we assume that you travel at a constant rate. For clarity we suppose you are trying to move from A to B where those points are ten miles apart.

First analysis: We first assume that to move from A to B you must take an endless number of steps, and each one requires the same amount of time (say one hour). We imagine that you do not linger at any of the halfway points, so after moving from one halfway point to the next you immediately move to the next one after that. Then the following chart illustrates the progress you would make with your first five steps:

STEP NUMBER	DISTANCE TRAVELED (in mi)	ELAPSED TIME (in hr)
1	5	1
2	$7\frac{1}{2}$	2
3	$8\frac{3}{4}$	3
4	$9\frac{3}{8}$	4
5	$9\frac{11}{16}$	5

What have we discovered? It will take you forever to complete your journey because each step takes one hour, and there are an endless number of them. To take your first 1 million steps will take 1 million hours, or just over 114 years. There is no limit to how much time your trip will take; for example, 1 million years after you start you will have taken more than 8.7 billion steps, so you would be very, very close to B, but not quite there yet.

Second analysis: If you are moving at the constant rate of five miles per hour, then the distance = rate × time formula shows that it will take you two hours to travel the ten miles from A to B. But let's dissect your motion into increments, and calculate how long it takes you to go halfway from A to B, then halfway from that position to B, and so on. To move from halfway point to halfway point you will have to travel 5 miles, then $2\frac{1}{2}$ miles, then $1\frac{1}{4}$ miles, and so forth. Since you are moving at the constant rate of five miles per hour, your movement over these increasingly shorter distances will take less and less time. The table below shows your progress over the first five intervals:

DISTANCE TRAVELED (in mi)	ELAPSED TIME (in hr)
5	1
$7\frac{1}{2}$	$1\frac{1}{2}$
$8\frac{3}{4}$	$1\frac{3}{4}$
$9\frac{3}{8}$	$1\frac{7}{8}$
$9\frac{11}{16}$	$1\frac{15}{16}$

Moving at the constant rate, your total elapsed time is increasing as you move from A to B, but it appears to be getting closer and closer to two hours (the time we would expect it to take you to move ten miles while traveling at five miles per hour). That is, of course, what happens, and the chart is misleading as to the nature of your motion; this was precisely Bergson's complaint (see chapter 3).

We now take these two views of motion back to the process of repeatedly doubling the number of sides of polygons inscribed in a circle and then carry it over to producing more and more, successively shorter line segments approximating the curve. If we assume that it takes the same amount of time for the polygon to morph from having a particular number of sides to twice as many sides, then we are in the situation described in our first analysis of motion. However, if we take the second point of view, that these successive polygons are produced in increasingly short intervals of time (just as God created angels in Gregory's example in chapter 4), then it is possible to produce an unlimited number of approximating polygons.

Recall the simplicity of Gregory's argument: Assume an arrow is shot at a target that is 10 feet away. While the arrow moves from the bow to the halfway point, 5 feet from the target, God inscribes a hexagon inside the circle. While the arrow moves from the position 5 feet from the target to the next halfway point, 2 ½ feet from the target, God doubles the number of sides of the inscribed hexagon and produces an inscribed twelve-sided figure. Then, while the arrow moves to the next halfway point, God doubles the number of sides of the polygon to obtain an inscribed twenty-four-sided figure. By the time the arrow reaches the target, it will have moved an unlimited number of shorter and shorter distances, and therefore, God will have created an unlimited number of inscribed polygons (with increasingly more sides). Although there was no last polygon created by God in this process, this sequence of polygons can be visualized as all being contained inside the circle and becoming better and better geometric approximations to it until the polygons and circle are virtually indistinguishable. (This also happens in the first analysis of motion, except that there is an intrinsic limit to how closely the polygons approximate the circle in our lifetime or in 1 million years. Just observing the process unfold we would not know if it was going to continue or simply stop after the appearance of the 1-millionth polygon. Of course this 1-millionth polygon and the circle would be indistinguishable to us. The advantage of the second approach is that all of these polygons will have been produced, so no matter what magnification we use, the polygons and circle will be indistinguishable.)

A mathematician would say that the sequence of polygons *converges* to the circle. What is central to the meaning of this phrase is that although no polygon ever equals the circle, the polygons are becoming better and better approximations of the circle and there is no limit as to how good these approximations can be. This is what also appears to be happening when we use more and more, shorter and shorter line segments to approximate the parabola. But how do we know that the *curve* at each stage, which consists of many short line segments, is geometrically approximating the original curve? To understand why these curves are indeed better and better approximations to the original curve, we need to look more closely at this idea of convergence, and instead of working with curves we first consider sequences of numbers.

CONVERGENCE

In the second table above, the total time that has elapsed as the body repeatedly moves halfway from A to B is, in hours, $1, 1\frac{1}{2}, 1\frac{3}{4}, 1\frac{7}{8}, 1\frac{15}{16}$, and so forth. Since the body moves through an unlimited number of halfway positions, there is no end to this listing of elapsed times. However, even though these numbers are becoming larger and larger, they are getting closer and closer to the value of 2, and there is no barrier between them and 2. Put differently, if we look at the differences, $2 - 1\frac{1}{2} = \frac{1}{2}, 2 - 1\frac{3}{4} = \frac{1}{4}, 2 - 1\frac{7}{8} = \frac{1}{8}$, we see that the differences between these numbers are becoming smaller and smaller, and they are approaching zero.

This concept involves two things: a sequence of numbers and a number they appear to be *converging* to. But there is another way to look at the sequence that can tell us if it converges to something without telling us what it converges to. It is an *existence criterion*. The intuition behind this criterion is that if a sequence is getting closer and closer to some number, then the numbers in the sequence must be getting closer and closer to each other. In the above example, the numbers are all getting closer and closer to 2, and the differences between successive terms, $1\frac{3}{4} - 1\frac{1}{2} = \frac{1}{4}$, $1\frac{7}{8} - 1\frac{3}{4} = \frac{1}{8}$, $1\frac{5}{16} - 1\frac{7}{8} = \frac{1}{16}$, and so forth, are getting closer and closer to zero. In this point of view, the numbers are converging to something because they are successively changing less and less. This means that we know the original numbers converge to something, without necessarily knowing what it is. This is a remarkable insight into the idea of convergence; it was made by the mathematician Augustin-Louis Cauchy (1789–1857), and convergence was the concept needed in the nineteenth century to provide the theoretical basis for Newton's version of calculus.

The convergence of a sequence of numbers is fairly easy to understand because it is possible to tell when two numbers are close to each other—their difference will be small, so approximately zero. Transferred back to the geometric realm, convergence is harder to describe, as there are no easily explained numbers we can associate with "the difference between two polygons" or "the difference between two curves." But Cauchy's idea applied in the geometric realm can help us out. We consider this idea in our approximations of both the circle and the parabola.

As the polygons inside the circle develop more and more sides, it is clear that once the number of sides is astronomically large, when the number of sides is doubled they change very little in shape. If you think of a point on one polygon having to move to a new position to become a point on the subsequent polygon, it is not too hard to imagine that the point does not have to move very far at all. Indeed, with the production of each successive polygon, the points will have to move less and less. As was the case with numbers, just knowing that the points on the successive polygons are closer and closer together demonstrates that these polygons have to converge to something.

We just happen to know, ahead of time, that these polygons converge to the circle. Another point of view is not to think of them as being contained inside the circle and getting closer and closer to it, but to think of this sequence as defining the circle. This is the concept of plenitude, which led de Cusa and Bruno to claim that there must exist infinitely many worlds, imported into mathematical thinking. Just knowing that the polygons form a convergent sequence of geometric objects, in that they are geometrically closer and closer together, means that they must converge to something. In this case, the object the polygons converge to is the circle; taking this point of view, the sequence of polygons defines the circle.

Similarly, as we use more and more line segments to estimate the length of the parabola, when we take one estimate and then obtain the next one by using more points, the line segments are increasingly close together. The points on the original line segments are very close to the points on the successive line segments. Using Cauchy's ideas, this means that the line segments converge to something—in this case the original parabola.

THE KOCH CURVE

One reason geometric convergence is such an important notion is that it permits mathematicians to study, and even produce, complicated geometric objects, and mathematicians can study these objects by examining the properties of a sequence of simpler geometric objects converging to the complicated object. Not all properties of the objects in the approximating sequence will carry over to the object they converge to. In Archimedes' example all of the converging objects have sides and corners while the circle they converge to is smooth, but the polygons' perimeters helped Archimedes understand the circumference of the circle.

To see these ideas at play we begin with a curve that was described by Niels Fabian Helge von Koch (1870–1924) in 1906, now called the *Koch curve*. We first have to confess that we cannot even draw a Koch curve, so complicated is it, but we can describe simpler geometric objects that converge to it. The process of constructing these simpler geometric objects is iteration of the following geometric process: Given a line segment, as on the top in Figure 11.2, replace it with the bent segment, on the bottom, which is obtained by replacing the middle third of the original line segment with two sides of an equilateral triangle.

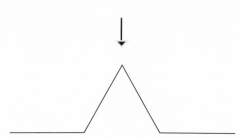

FIGURE 11.2. The first step in developing the Koch curve is to replace the middle third of a line segment with a bent segment consisting of two segments, each the length of the removed middle third.

The next curve is obtained by then replacing the middle third of each of the four straight segments in the second figure in Figure 11.2 by two sides of an equilateral triangle.

FIGURE 11.3. The process applied to the line segment in Figure 11.2 is then applied to each of the four segments in the bent segment.

If this process is repeated two more times we get the configurations in Figure 11.4.

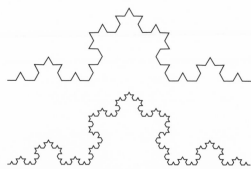

FIGURE 11.4. Repeating the process of replacing each straight segment with a bent segment leads to successively more jagged curves.

It is fairly clear that the curve is more jagged after each iteration, but its points do not move too much from one iteration to the next. In fact the difference between the fiftieth and fifty-first curve cannot be seen. Using our naive understanding of convergence, and appealing to mathematical plenitude, this sequence of curves will converge to something—that something is the Koch curve.

An important observation to make is that in each step a straight segment is replaced by a bent segment that will be four-thirds as long (because the middle third is replaced by two pieces of that length). This tells us something about the Koch curve—it does not have a finite length, because if the original segment's length is one unit, then the successive curves have lengths:

second curve: $4/3 = 1.333 \ldots$
third curve: $16/9 = 1.777 \ldots$
fourth curve: $64/27 = 2.370 \ldots$
fifth curve: $256/81 = 3.1604 \ldots$
sixth curve: $1024/243 = 4.213 \ldots$

It is possible to show that these numbers are simply becoming larger and larger without any bound; they do not form a convergent sequence. And although the Koch curve has an infinite length (i.e., a nonfinite length), it is contained in the small one-by-one square constructed on the original segment. So it is bounded and infinite. This means that the Koch curve must not look like any curve we have previously seen. The Koch curve is so jagged that once we have developed a more formal definition of the dimension of a geometric object than we used in chapter 8, we will see that it is not even one-dimensional. To understand this last statement, we reconsider what we mean by the dimension of a geometric object.

NEW WAYS TO THINK ABOUT DIMENSION

The dimension of an object can be described in two ways that will make it possible to calculate the dimension of the Koch curve. Each of these notions of dimension relies on a process. One is based on subdividing the edges of the object in order to produce more (albeit smaller) copies of the same shape, and the other is to think about gluing copies of the object together to get a single larger copy of the same object. These two approaches lead to the same number for the dimension of an object, but the second is slightly easier to visualize so we explain it here.

In order to calculate the dimension of an object using this approach, we have to answer the following, basic question: How many copies of a given object are required to produce a larger object of the same type? There is some ambiguity in the terms used in this question, so we first consider the simplest possible example, a line segment that is one inch long, and ask, how many copies of this line segment does it take to make a line segment twice as long or three times as long? The answer is that two copies of this segment can be used to produce a line segment twice as long and three copies of this segment can be used to produce a line segment three times as long.

We can apply these same ideas to a square. We begin with a small square, for example a square that is one inch by one inch. We need to be clear about what it means for a square to be twice as large, or three times as large, as a given square. By this we do not mean a square having twice the area, or three times the area, of the original square; we mean twice as large, or three times as large, in the sense of magnification. It will take four squares to produce a square twice as large, in this sense, and nine squares to produce a square three times as large:

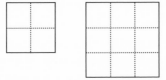

FIGURE 11.5. Four small squares of equal size may be used to construct a square two times as large; nine small squares of equal size may be used to construct a square three times as large.

Next let's see how these constructions reflect what we already believe—that a line segment is one-dimensional and that a square is two-dimensional. For the line segment we had the correspondence:

MAGNIFICATION FACTOR	NUMBER OF COPIES NEEDED
2	2
3	3

For the square we discovered the correspondence:

MAGNIFICATION FACTOR	NUMBER OF COPIES NEEDED
2	4
3	9

The dimensions of the line segment and square can then be uncovered from an equation that formally defines the dimension of an object:

$$(\text{magnification factor})^{\text{dimension}} = $$
number of copies of objects needed

Applying this formula to the data we discovered for the line segment and the square, we see that, as expected, the line segment is one-dimensional and the square is two-dimensional. (Note that this concept of dimension can only be applied to objects that can be constructed out of smaller versions of the same object; it cannot be applied to all curves or to objects like circles.)

We could also use this formula to rediscover that the dimension of the cube is three, but instead we move on to an attempt to use it to calculate the dimension of the Koch curve. Although the Koch curve is defined as the curve obtained through completing the infinite process described above, we can examine what happens at each step of that process to develop an understanding of the dimension of the completed object.

We start with a simple part of the Koch curve, the bent segment in Figure 11.2, and ask how many copies of this basic piece would it take to make a larger piece? To think about the meaning of this question, look back at the process by which the geometric approximations of the Koch curve were formed. The next curve in the iteration process can be imagined to be made up of four copies of the original Koch-segment, and this curve is three times as large as the original curve, since it is three times as wide and three times as tall. Notice that it is not possible to use copies of the original segment to produce a curve of the same type that is twice as large; the process of producing a larger curve always requires replacing each segment with four segments. Thus when the magnification factor is three we require four copies of the original segment, and so the dimension of the Koch curve satisfies the equation:

$$3^{\text{dimension of Koch curve}} = 4$$

It follows that the dimension of the Koch curve is an unending decimal that begins: 1.26185. This means that the Koch curve is so jagged that its dimension is larger than one, but less than two.

SNOWFLAKES, COASTLINES, AND JACKSON POLLACK

Clouds are not spheres, mountains are not cones, coastlines are not circles, and bark is not smooth, nor does lightning travel in a straight line.

— Benoit Mandelbrot, The Fractal Geometry of Nature *(1982)*

In 1967 the mathematician Benoit Mandelbrot (b. 1924) published a paper with the provocative title "How Long Is the Coast of Great Britain: Statistical Self Similarity and Fractional Dimension," wherein he suggested that jagged curves, like the Koch curve, might indeed appear in nature. Mandelbrot did not suggest that such regularity would be discovered, but that under increased magnification, something like the coastline of Britain would simply look more and more jagged, as does the Koch curve. The measurement of the coastline of Britain would depend on the length of the ruler used. If the ruler is too long, say one mile, then it will not capture the indentations of small coves. The shorter the ruler used, the larger and more accurate the measurement. Below are the different measurements one would obtain for the length of the coastline of Britain if one were to use measuring sticks of different lengths.

RULER LENGTH (in km)	LENGTH OF COASTLINE (in km)
500	2,600
100	3,800
54	5,770
17	8,640[2]

We can extrapolate from these data that for rulers shorter than 17 kilometers the coastline's length would be even longer.

The complexity of the coastline of Great Britain cannot be determined by examining a sequence of increasing lengths; dimension measures its complexity. But the notion of dimension discussed above is only applicable to mathematically generated fractals, not to existing, material curves. However, there is a way to define the dimension of a fractal, which relies on the fractal's being a very complex curve with an endless number of twists and turns that can be experimentally determined.

If we imagine looking at the Koch curve through a microscope, it looks jagged whether the magnification scale is ten, one hundred, or Jimmy Carter's "thousand billion." The same holds, to a degree, with some physical *curves*, such as the edge of a snowflake or the coastline of Great Britain. The motivation for this third description of dimension of a complicated curve is that while it agrees with the previous calculation of the dimension of a mathematical fractal, it provides a way to measure the dimension of a leaf or a Jackson Pollack painting.

Briefly, this dimension depends on how jagged the curve is when it is viewed under greater and greater magnification. The idea behind this concept of dimension, and we are only going to give the idea to indicate that it is something that could be calculated for a physical object, is to begin by covering the curve with a square grid. Then the curve will pass through some but not necessarily all of the squares in the gird. When smaller squares are used to make the grid, the curve will again pass through some of the squares but not all of them. If this process is repeated over and over, using smaller and smaller squares, a relationship will emerge between the number of squares the curve passes through and the size of the squares. The dimension of this curve will be an exponent in this relationship. The grid-dimension of the Koch curve is 1.269 (close to the precise value obtained above); the grid-dimension of an oak leaf is 1.7.

In 1999 Richard Taylor, Adam Micolich, and David Jonas published an article titled "Fractal Expressionism." The researchers applied this grid method not to mathematical curves or to geological formations but to some paintings of Jackson Pollack (1912–56). Pollack's paintings do not have fractal-like properties because they are abstract but because of Pollack's painting method. As Taylor, Micolich, and Jonas wrote, "There were two revolutionary aspects to the way in which Pollack painted and both have the potential to introduce chaos" and lead to fractal patterns. The first was how Pollack positioned himself in relation to the canvas. Pollack placed the blank canvas in a horizontal position and he moved his body over the canvas—sometimes suspended above the canvas by a harness that allowed him to freely move in different directions. Pollack's body moved like a leaf floating on a river whose position changes in response to slight variations in conditions. The second "revolutionary aspect" of Pollack's method was not entirely revolutionary, but it was unorthodox. Pollack did not use brush strokes but dripped the paint onto the canvas in a continuous flow. This introduced another chaotic element into Pollack's painting, that of fluid flow.[3]

Using the grid method the article's authors were able to demonstrate that, over time, Pollack apparently refined the fluidity his method allowed, and the fractal dimensions of his paintings increased.

PAINTING	DATE	FRACTAL DIMENSION
Untitled: Composition with Pouring II	1943	about 1
Number 14	1948	1.45
Autumn Rhythm: Number 30	1950	1.67
Blue Poles: Number 11	1952	1.72[4]

Although you cannot determine the fractal dimension of a curve just by looking at it, Pollack's paintings do seem to become increasingly complex over time.

INFINITY

Earlier in this chapter, the Koch curve was expressed as the limit of a convergent sequence of geometric curves. And each of these curves had to be constructed, so for the Koch curve to have any meaning we have to be able to construct, or imagine constructing, the unlimited number of geometric objects that converge to it. Of course this cannot be done physically. But by imagining these constructions being completed over the increasingly shorter time intervals, as an object moves from one halfway point to another halfway point in traveling from A to B, we can at least have an idea of how this process could be completed.

The acceptance of the completion of certain infinite processes is central to the modern theory of mathematical infinity. What is at stake is how to answer the question

> What does it mean to say that a collection of
> mathematical objects is infinite?

We already know Aristotle's response, a collection is infinite if no matter how much of the collection has been contained or described, some if it will have been left out. According to this definition we have already encountered several collections that are potentially infinite by addition, for example the counting numbers, 1, 2, 3, 4, . . . and the even counting numbers 2, 4, 6, 8. . . . We can describe other endless collections, such as the collection of all decimals consisting only of nines after their decimal point, .9, .99, .999, and so forth, or of all decimals that start with a nine after the decimal point, .91, .997, .9084, and so forth. But if we imagine these collections as being completed, then we run up against the paradoxes of having greater and lesser infinities. In the fourteenth century Gregory of Rimini saw no difficulties with this containment of one infinite collection inside another, but before the late nineteenth century most mathematicians were at least uncomfortable with these violations of the whole is greater than the part.

Georg Cantor (1845–1919) overcame the paradoxes following from the assumption that the whole is greater than the part, by appealing to the completion of infinite processes to develop a convincing theory of mathematical infinity. It was Cantor's theory that led to Russell's allusion to infinite numbers at the beginning of this chapter. Central to Cantor's theory is a reevaluation of what it means to *count* objects.

There is an important, almost hidden, idea in our calling the numbers 1, 2, 3, . . . the *counting* numbers that was not fully appreciated until the nineteenth century. This is one of the many examples in the history of ideas where some important principle was not noticed until someone pointed it out, and once the principle has been articulated, it becomes difficult to understand why it was not always understood. Ideas of this sort can prove to be the most fertile of all, and the one implicit in our notion of counting has led to a satisfactory understanding of the mathematically infinite.

We begin with an imaginary tale. Suppose that you are assigned the task of counting the number of animals in a room. As soon as you walk into the room you realize that it is swarming with cats. There are cats crawling over all of the furniture, swinging from the drapes, and chasing each other in furious circles around the room's perimeter. So you start counting, the orange tabby sleeping in the chandelier is cat number 1; the long-haired black cat staring into the corner is cat number 2; the two white cats chasing each other up and down the drapes are cats number 3 and 4. By the time you have reached cat number 17, another orange tabby, you realize that cat number 1, the orange tabby that had been sleeping in the chandelier, has disappeared. So you don't know if cat number 1 and cat number 17 are the same orange tabby or not, and since there are easily a dozen orange tabbies in the room, you decide to take another approach.

You leave the room, fighting off the cats trying to get out, and retrieve a roll of masking tape and a marking pen. You then return to the room with your tape and pen, determined to obtain an accurate count of the cats in the room. You tear off a piece of tape, write a 1 on it and stick it to the head of the cat that is already sleeping on your foot. Then you peel off another piece of tape, write a 2 on it, and stick it to the head of another cat. You continue this process, labeling cats 3, then 4, and so forth. Amazingly, not one of the cats tries to remove the tape from its head. You finally tape the number 56 to the black cat inside the empty aquarium and look around the room. Every cat has a piece of tape on its head. You know that you were careful neither to skip any number nor to use any number twice when you were writing numbers on the pieces of tape. Using the hidden idea, of which you are totally unaware, you conclude that there are 56 cats in the room.

The hidden idea is what mathematicians have come to call a *one-to-one correspondence*. We will elaborate on this concept when we formalize our definition of a mathematically infinite collection, but for now let's just see where it is hidden in your cat-counting nightmare. What you have effectively done by placing the consecutively numbered pieces of tape on the cats' heads is to find a way to associate the list of numbers 1 through 56 with the collection of cats in such a way that each number corresponds to one and only one cat and no cat has been missed. Such an association between the collection of numbers and the collection of cats is a one-to-one correspondence between the collections. To understand the power of this simple concept in understanding quantity abstractly, let's briefly return to your attempt to count the cats.

Suppose that when you leave the room to retrieve the tape and marking pen, you can only find tape. This means that you cannot write any numbers on the pieces of tape. It is still possible to count the cats in the room; it will simply take an additional step. Upon reentering the room, you tear a piece of tape from the roll and stick it onto a randomly chosen cat's head. You then tear another piece of tape from the roll and stick it onto another cat's head, and then you stick a piece of tape to yet another cat's head. You continue this process until every cat has a piece of tape stuck to its head. Satisfied that you did not miss any cats, you reverse this process and remove the pieces of tape from each and every cat's head and leave the room holding all of the pieces. You clearly do not know how many cats are in the room, but you do know that there is a one-to-one correspondence between the cats in the room and the pieces of tape in your hand. (This is because every piece of tape corresponds to one and only one cat, and every cat corresponds to one and only one piece of tape.) If you want to determine the number of cats in the room, you simply need to count the pieces of tape you have. In other words, to find out how many cats there are in the room, you do not have to count the cats; you only need to count the pieces of tape.

The significance of establishing a one-to-one correspondence between the items in two collections is that it is possible to conclude that the two collections are the same size without having to assign a number to these quantities. This is precisely the idea needed to compare infinities.

SET THEORY

We are now able to reconsider an earlier paradox of counting. If we start to count the even counting numbers, 2, 4, 6, and so forth, the beginning of our count looks as follows:

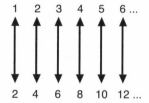

FIGURE 11.6. A pairing between the counting numbers and the even counting numbers.

This was seen to be a paradox because it appears to indicate that there are just as many even counting numbers as there are counting numbers. And as the collection of even numbers is contained in the collection of counting numbers, this correspondence appears to violate the fundamental principle that the whole is greater than the part.

But if we want to admit the existence of infinite collections into mathematics, then the correspondence above violates the extension of the principle the whole is greater than the part to infinite collections, that there cannot exist a greater and a lesser infinity. Thus, following Aristotle, we could deftly dispose of this paradox by allowing only potentially infinite collections, even in the world of mathematical objects and relationships. Still respecting Aristotelian ideas, we restate Aristotle's definition of a collection being *potentially infinite by addition* using the concept of a one-to-one correspondence as:

> A collection is potentially infinite by addition if it cannot be put into a one-to-one correspondence with any finite collection of counting numbers, such as $1, 2, \ldots, 7$ or $1, 2, \ldots, 981$.

But such a definition does not tell us anything new. It simply says that a collection is potentially infinite by addition if it is unending. The way to give a positive formulation for a collection to be infinite, and not just potentially infinite, is to allow for the completion of infinite processes. This means that we want to view the above correspondence, between the counting numbers and the even counting numbers, as a completed pairing. And to arrive at this, we need a somewhat more formal notion of a collection.

One of the most significant conceptual leaps in the history of mathematics was taken in the nineteenth century when mathematicians gave a precise definition of a *collection of objects*, and so provided the language for discussing *the whole* and *the part*, which Gregory of Rimini did not provide. What mathematicians uncovered is the concept of a *set*; a concept so all-encompassing that it led many mathematicians and philosophers to declare that mathematics is nothing but the study of sets. The concept of a set is both straightforward and ultimately paradoxical. The most common definition is that a "set is a collection of objects."

Cantor used the concept of a set in order to make sense of the mathematically infinite, and Cantor defined a set as "any collection into a whole M of definite and separate objects m of our intuition or thought."[5] Implicit in Cantor's vague but more useful definition of a set is that if we are, somehow, presented with a set, M, and an object, m, we must be able to determine whether or not m is one of the objects in M. This is the law of the excluded middle in disguise. If our set M is defined to be the collection of all objects that satisfy some property P, then M is only unambiguously described if, for any object under consideration,

P either holds for the object, in which case that object is in M, or P does not hold for the object, in which case the object is not in M. This conception of a set is problematic, ultimately leading to paradoxes that undermined the late nineteenth-century and early twentieth-century hopes to reduce all of mathematics to logic and set theory. But some version of the idea that a set is a collection of objects, which is defined in such a way that we can always tell whether or not a particular object is in the set, remains.

The concept of a set allows us to accomplish what Augustine said only God could do and what we based on process in the last chapter—it permits us to see, or at least speak of, an infinite collection as a completed whole. We can think of the counting numbers 1, 2, 3, 4 . . . as an endless list, or we can imagine collecting them all together into a single set. We simply let W be the set whose elements are all of the counting numbers. W contains nothing more and nothing less than all of the counting numbers. The question of whether or not W has been completed is meaningless; if an object is a counting number it is in W, and if an object is not a counting number it is not in W. The set W has the same sort of existence of any line segment, angle, or pair of parallel lines.

A DEFINITION OF THE MATHEMATICALLY INFINITE

Using the language of sets we can now give a provisional, positive definition of an infinite set:

> A set is infinite if it can be put into a one-to-one correspondence with the set of all counting numbers.

This definition of an infinite set needs to be modified, but it is an important first step. What is so important about this definition is that it takes the existence of the one-to-one correspondence between the counting numbers and the even counting numbers, which has been seen to be paradoxical, as the justification of the claim that the set of all even counting numbers is infinite.

If this were the only progress Cantor made toward understanding the mathematically infinite, we would not still be discussing his work well over a century later. All we have done so far is to simply extend Aristotle's definition of potentially infinite by addition to infinite by allowing ourselves to imagine that we have completed the listing of all of the elements in a particular collection. But Cantor did more than this with his theory of sets and use of one-to-one correspondences, and to understand Cantor's achievement, we return to the one-to-one correspondence between the collection of all counting numbers and the collection of even counting numbers.

We viewed this correspondence as a process of sequentially associating the counting numbers with the even counting numbers and imagined that this process could be completed. All that we needed was an unambiguous way to know which counting number corresponds to which even counting number. The association could also be made mathematically formal: A counting number n corresponds to an even counting number $2 \times n$. Having this rule would allow us to represent this correspondence as

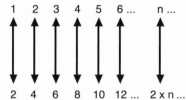

FIGURE 11.7. A mathematical description of the one-to-one correspondence between all of the counting numbers and all of the even counting numbers.

But some one-to-one correspondences are harder to describe by a mathematical rule, so we rely on our intuition that the correspondence is unambiguous and could, if necessary, be continued indefinitely. A few examples will suffice to illustrate this point.

Let's start with the set obtained by taking the counting numbers 1, 2, 3, ... and adding in 0, so the set is 0, 1, 2, 3, 4.... This is clearly an infinite set, but can we prove it is infinite by finding a one-to-one correspondence between this new set and the set of all counting numbers? We start with a list of all counting numbers above a list of the numbers 0, 1, 2....

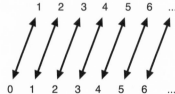

FIGURE 11.8. A one-to-one correspondence between the collection of all counting numbers and the collection of all counting numbers with zero included.

The arrows describe a one-to-one correspondence between these two collections, and so, according to the definition of an infinite set, the set containing the numbers 0, 1, 2, 3, . . . is an infinite set.

It is possible to adopt this visualization idea to show that the set of all integers, . . . , −2, −1, 0, 1, 2, . . . and the set of counting numbers can be put into a one-to-one correspondence. The idea here is slightly different in that we do not list both collections, but only the collection of all positive and negative integers. We then describe how to systematically associate them with the counting numbers. We begin by listing the integers in two rows, and we count them by following the arrows (so, for example, 0 is the first integer, 1 is the second integer, and −1 is the third integer):

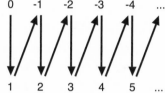

FIGURE 11.9. A systematic way to count the collection of all positive and negative whole numbers. The count follows the arrows: zero is first; one is second; minus one is third, and so forth.

Following the arrows yields the correspondence

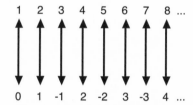

FIGURE 11.10. The one-to-one correspondence between the counting numbers and the collection of all positive and negative whole numbers emerging from the counting process in Figure 11.9.

A subtler listing shows that a one-to-one correspondence can also be found between the collection of counting numbers and the collection of all fractions. Since the collection of all fractions is clearly an infinite one, our definition of an infinite set—as a set having a one-to-one correspondence with the collection of counting numbers—seems to be sound. However, the next example reveals our provisional definition's shortcomings.

We consider the set of all decimals between 0 and 1, for example .2, .406, and .2626262626. . . . This is clearly an unending set, so it should be infinite, according to our definition of an infinite collection above. So assume that there is a one-to-one correspondence between all of these decimals and the counting numbers. For example, this correspondence might start out as

1. ⟷ .783656392758
2. ⟷ .13131313 . . .
3. ⟷ .7362527239373
4. ⟷ .00000000000000000001

FIGURE 11.11. The beginning of a possible counting of all decimals between 0 and 1.

We claim that *any* such list must miss some decimal between 0 and 1. Here is the argument: Suppose we have a listing of all of the numbers between 0 and 1, perhaps beginning as the above list does. It is then possible to describe a number, β, which is between 0 and 1, and cannot possibly be on the list. We describe β by successively describing each of its decimal digits.

Let β have its first decimal digit be different from the first decimal digit of the number labeled 1, so the first decimal digit of β is something other than 7. Let it be 8, so β begins: $\beta = .8$. Next add a second decimal digit to β so it cannot possibly equal the second number on the supposedly complete list. This means that we need to choose the second decimal digit of β to be something other than the second decimal digit of the decimal numbered 2, so something other than 3, for example, 7, so β now has two decimal digits: $\beta = .87$. Next choose the third decimal digit of β to be something other than the third decimal digit of the decimal numbered 3, so something other than 6, for example, 5. So β's decimal expansion starts out $\beta = .875$, which means that it cannot be equal to any of the first three numbers on our list. Continue to choose the decimal digits of β by the rule. Choose β's kth decimal digit so that it is not equal to the kth decimal digit of the number on our list that is numbered k. Assuming the completion of this infinite process, we will have found a number β not on the list.

This process works no matter whatever the initial list looks like. Thus, no such list can ever contain all of the decimals between 0 and 1. There are two possible conclusions:

1. the collection of decimals between 0 and 1 is not an infinite set, or
2. our definition of an infinite set is not sufficient.

Since we would like to let any nonfinite set be infinite (since any non-finite set is infinite in the Aristotelian sense), the first possibility must be rejected. Thus we need to reexamine our provisional definition of an infinite set to obtain one that shows that both the collection of all even counting numbers and the collection of decimals between 0 and 1 are infinite sets.

A PARADOXICAL DEFINITION OF AN INFINITE SET

Using an idea going back to Gregory of Rimini, we say that a set is infinite if it seems to violate the principle that the whole is greater than the part. In our language:

> A set is infinite if it can be put into a one-to-one correspondence with a part of itself.

According to this definition, if we can concoct a seemingly paradoxical one-to-one correspondence between a set and a collection contained in the set, as we did between the counting numbers and the even counting numbers, then the set must be infinite. Clearly under this definition the counting numbers are infinite. Though we will not prove this next statement, there is a one-to-one correspondence between the decimals between 0 and 1 and the collection of all real numbers. Thus, under this definition of an infinite set, the collection of all real numbers is infinite.

Cantor showed that this definition leads to different sizes of infinities. We only describe two of them. To begin with there are all of those infinite sets that can be put into a one-to-one correspondence with the set of counting numbers. Yet there are many entirely ordinary infinite sets that cannot be put into a one-to-one correspondence with the counting numbers; many of these seem to have a one-to-one correspondence with the real numbers. So these two sizes of infinity are the size of the collection of all counting numbers and the size of the collection of all real numbers.

A single symbol, such as ∞ obscures this difference, so Cantor introduced two new infinity symbols, $\aleph_0$ and c, for the first and second of these infinities, respectively. These symbols are not numbers in our earlier senses: They are neither geometric magnitudes nor algebraic numbers. These are precisely two of the infinite numbers Russell referred to in the quote that opened this chapter. Russell's reason for calling these entities numbers is based entirely on their utility; just as 7 represents the number of days in the week, and 1,000,000,000,000 (1,000 billion) the possible number of stars in Jimmy Carter's cosmos: $\aleph_0$ represents the number of counting numbers and c the number of real numbers.

If we accept $\aleph_0$ and c as numbers, then Russell's claim that it is not true "that a number is increased by adding one to it" makes sense. Looking back at the counting numbers, we saw that when we added one more number to the set (we added 0), the new and old set can be put into a one-to-one correspondence with each other. Thus, they both have a *cardinality* of $\aleph_0$. Similarly adding another element to the real numbers, such as $\sqrt{-1}$, will not change their cardinality one iota.

Cantor went further than the collection of counting numbers, giving rise to $\aleph_0$, and the collection of real numbers, giving rise to c. He proved that there is an unending tower of larger and larger infinities. In this sense, the counting numbers are the smallest infinite set. We will not explain how Cantor moved beyond the real numbers, but once he described this process, he produced *larger* and *larger* infinities simply by iterating it. The hierarchy of infinities is endless.

POSTSCRIPT

[It] is just as impossible to get to essence by accumulating accidents as to reach 1 by adding figures to the right of 0.99.

— *Sartre*, The Emotions: Outline of a Theory *(1948)*

Jean-Paul Sartre's claim that it is impossible to "reach 1" by appending more nines to the decimal .99 or .999999 or even .999999999999 is correct, because implicit in his assertion is the assumption that someone, or something, will physically write each of those nines. It seems reasonable to extrapolate from this interpretation that Sartre (1905–80) is correct and that it is impossible to ever "reach 1" using nines. This extrapolation is both correct and incorrect, and this dichotomy depends on whether we reject or accept the completion of certain infinite processes. If we are going to allow for the completion of infinite processes, then it is fair to try to understand an unending decimal, such as .999 ..., where there is an unending sequence of 9s to the right of the decimal point. Before we address this, let us reconsider what we even mean by a decimal.

Consider the decimal .234, which seems simple enough. To understand the meaning of this decimal, in Greek terms, we need to see it as a ratio of two whole numbers. Since each position after the decimal point has a special meaning, it is possible to translate the symbols .234 back into the standard fraction notation. In any decimal expansion of a number, the first position after the decimal point represents a number of one-tenths; the second position after the decimal point represents a number of one-hundredths; the third position after the decimal point represents the number of one-thousandths; and so on. Thus .234 can be expressed as a sum of three fractions:

$$.234 = (2 \times \tfrac{1}{10}) + (3 \times \tfrac{1}{100}) + (4 \times \tfrac{1}{1,000}) = \tfrac{2}{10} + \tfrac{3}{100} + \tfrac{4}{1,000}.$$

A common denominator for these three fractions is 1,000 so $.234 = \tfrac{234}{1,000}$.

The reverse process, of beginning with a fraction and finding a decimal that represents it, might seem to be a daunting task. What it entails is starting with a fraction, such as $\tfrac{21}{50}$ and expressing it as a sum of fractions whose denominators are 10, 100, 1,000, 10,000, and so forth. Fortunately this is a relatively easy process; it involves performing a division of the denominator of the fraction into the numerator. This process is not important here, but what it implies is important. The fraction $\tfrac{21}{50} = .42$, a nice, finite decimal, because

$$\tfrac{21}{50} = \tfrac{42}{100} = \tfrac{40}{100} + \tfrac{2}{100} = \tfrac{4}{100} + \tfrac{2}{100} = .42.$$

But not all numbers are so easily expressed as a decimal because not all fractions are equivalent to a fraction whose denominator is a power of ten. For example we cannot have any equality of fractions $\tfrac{1}{3} = \tfrac{N}{1,000,000}$, or any other denominator that is a 1 followed only by zeros. This means that $\tfrac{1}{3}$ cannot be expressed as a sum of a few fractions whose denominators are powers of ten. If we try the long-division approach, dividing the denominator into the numerator, we discover that the decimal for $\tfrac{1}{3}$ begins .33333 and the process never ends. Thus, the decimal expansion for $\tfrac{1}{3}$ would have to have an unlimited number of 3s after the decimal point. If we want to have representations for all geometric magnitudes, we have to accept as a decimal something we cannot even write down: decimals with an unlimited number of nonzero digits. Thus, we are left with the following dilemma concerning Stevin's decimal notation, which further argues for the acceptance of complete infinite processes: If we want all simple fractions to have decimal representations we have to accept the completion of infinite processes. And, in a slight digression, we explain why accepting unending decimals implies that $1 = .99999999\ldots$.

There are several ways to see that 1 and $.999\ldots$, where in every position past the decimal point we have a 9, represent the same length. Perhaps the most direct way is to use the idea that if two numbers are not equal then there must be some number between them, for example their average. (If we conceive of numbers as being points on the real line this is the statement that the continuum is infinite by division.) So we assume that $.999\ldots$ and 1 are different and then take an arbitrary number between them, which we denote by t:

$$.9999\ldots < t < 1$$

The idea is to try to understand what the decimal representation for t must look like. Since $.9 < .999\ldots$ and $.999\ldots < t$, we know that: $.9 < t$ and so $.9 < t < 1$. This inequality tells us that the first number after the decimal place in t will be a 9, so $t = .9xxxxxx$. Next, we use the fact that $.99 < .999\ldots$ to obtain: $.99 < t < 1$, and so the second number after the decimal point in t is also a 9, i.e., $t = .99xxxxx$. Continuing in this way we may repeat this process to find that the third decimal digit of t is a 9, and the fourth decimal digit of t is 9, and as far as we go each of t's decimal digits will be a 9. Stated slightly differently: t is a decimal in which every decimal digit is a 9. Thus t and $.999\ldots$ are indistinguishable, so there is no such t strictly between 1 and $.999\ldots$. That is, these two numbers must be equal.

12

Elegance and Truth

The year is 2002 and here we are at a symposium on Foundations and the Ontological Quest. The first thing to say is how bleak the present situation is. In foundational studies of mathematics . . . we have been stuck for seventy years; despite numerous books, articles, and meetings there has been no real progress.
— *Edward Nelson, "Syntax and Semantics" (2002)*

As mathematicians realized that geometry is just as likely to be non-Euclidean as Euclidean, that numbers could transcend both algebraic and geometric definitions, that there is an unending hierarchy of mathematical infinities, and that a curve could be bounded and have infinite length and even have a dimension that defies intuition, they began to reexamine the most fundamental aspects of their enterprise. Late in the nineteenth century and early in the twentieth century, mathematicians again sought to answer a question that had plagued the Pythagoreans, medieval atomists, Gregory of Rimini, and Gauss: What are mathematical truths, and how do we find them?

We begin with a more formal version of the first part of this question:

What does it mean for a mathematical statement to be true?

The most easily understood definition of the truth of a statement is one derived from Aristotle:

The truth of a sentence consists in its agreement with reality.

The sentence "the moon is made of cheese" is true if it happens to be the case that the moon is made of cheese. This statement's truth can be empirically verified, at least in theory. Transferring this naive notion of truth into the realm of mathematics presented mathematicians with a serious difficulty. It is not clear that there is a reality against which a statement can be compared, and if there is such a reality it may not be immediately accessible.

The twentieth-century philosopher W. V. Quine (1908–2000) began his influential article on ontology "On What There Is" by considering the question, "What is there?" This question covers a multitude of special cases, each of which examines a different aspect of reality or faith. Quine knew that each of the questions Do angels exist? Do unicorns exist? Do trees exist? and Do parallel lines exist? presupposes a different sort of *existence*. But Quine's answer to his initial question covered all of these cases, and was both absolutely correct and not at all enlightening; Quine wrote, this question "can be answered . . . in a word—'Everything'—and everyone will accept this answer as true. However . . . there remains room for disagreement over cases."[1]

Mathematicians have not always given the same answer to the questions "do mathematical objects exist" or the assumption-laden "which mathematical objects exist?" They have certainly disagreed over cases. This is not the place to enter into a detailed discussion of the philosophy of mathematics, but in order to understand the modern conception of mathematical beauty and mathematical truth, a certain familiarity with three early twentieth-century philosophies of mathematics is informative. Put simply, these three views of mathematics can be distinguished by how they answer the question, which mathematical objects exist? The three answers, which are the basis for these philosophies, are all of them; some of them; and none of them. The mathematicians who would have given these answers are known, respectively, as *logicists*, *intuitionists*, and *formalists*.

Later in his article, Quine wrote, "Classical mathematics . . . is up to its neck in commitments to an ontology of abstract entities," and he gave the example of a prime number greater than one million.[2] Both the logicists and intuitionists believed such a prime number existed, but that is just about the limit of their agreement. The *logicists*, as represented by Bertrand Russell and Alfred Whitehead (1861–1947), believed there was a Platonic realm in which mathematical objects exist; the *intuitionists*, as represented by the Dutch mathematician L. E. J. Brouwer (1881–1966), believed that the integers were the only mathematical objects whose existence was assured. As the mathematician Leopold Kronecker (1823–91) is widely quoted as having said, "God made the whole numbers, all the rest is the work of man."[3] (Kronecker so fiercely held this philosophy that he vigorously opposed Cantor's attempt to obtain a teaching position at the University of Berlin because Cantor's theory of infinite sets required the completion of infinite processes.)

The *formalist* point of view was decidedly not Platonic. The founder of this school was David Hilbert (1862–1943) who has been quoted as saying that mathematics "is nothing more than a game played according to certain simple rules with meaningless marks on paper."[4] The formalist point of view was that the game being played was the deduction of theorems from axioms, without regard to the content of the axioms. We may, if we so choose, give mathematical meaning to the symbols used, but that is not necessary.

With this brief overview of the three main schools of mathematical thought we return to the original question:

What does it mean for a mathematical statement to be true?

For mathematicians believing there is a Platonic realm of mathematical existence for all mathematical objects, mathematics is *discovered*, so there is a simple definition of mathematical truth based on Aristotle's:

A mathematical statement is true if it
agrees with the Platonic reality.

With this definition, truth can be found, at least in principle, in two ways—either through revelation or analysis.

But for both the intuitionists and formalists, mathematics is not discovered so much as it is *invented*. For the intuitionists, this invention is based on constructing new objects from the integers, for example the rational numbers or algebraic numbers; for the formalists, this invention is based on the ideas we attach to our symbols. For both intuitionists and formalists, the Aristotelian definition of truth is inappropriate. And having been misled by their intuitions about parallel lines and self-evident truths such as "the whole is greater than the part," all mathematicians were hesitant to accept *revealed* truths, or at least hesitant to admit that they accepted revealed truths. So mathematicians had a choice—they could, as some chose to do, endlessly debate the existence, or nonexistence, of mathematical objects, or they could settle on an operational definition of mathematical truth. The definition of truth they settled on is the same one that is implicit in Euclid's presentation of geometry:

> A mathematical statement is true if it can be derived (using commonly accepted principles of argumentation) from mathematical statements that are already known to be true.

Although mathematicians might quibble about which statements are "already known to be true" and about the "commonly accepted principles of argumentation," they generally accept this operational definition of a mathematical truth.

Alas, there are several intrinsic, logical difficulties with this definition of mathematical truth, one concerning the rules of logic that are among the "commonly accepted principles of argumentation" and another concerning the nature of all axiomatic, deductive systems. The first of these could be said to be *linguistic*: It is the problem of formalizing deductive rules. Indeed this difficulty was part of what drove Russell, Whitehead, and Hilbert to formalize the rules of inference. An example of this difficulty was illustrated in Borges' "Avatars of the Tortoise," where he attributed it to Lewis Carroll (1832–98). Borges wrote that at the end of their "interminable" race, in Zeno's paradox Achilles and the Tortoise, the two competitors begin to discuss geometry:

> They study this lucid reasoning:
>
> a) Two things equal to a third are equal to one another.
> b) The two sides of this triangle are equal to MN.
> c) The two sides of this triangle are equal to one another.
>
> The tortoise accepts the premises a and b, but denies that they justify the conclusion. He has Achilles interpolate a hypothetical proposition:
>
> a) Two things equal to a third are equal to one another.
> b) The two sides of this triangle are equal to MN.
> c) If a and b are valid, z is valid.
> z) The two sides of this triangle are equal to one other.
>
> Having made this brief clarification, the tortoise accepts the validity of a, b and c, but not of z. Achilles, indignant, interpolates:

d) if a, b and c are valid, z is valid.

And then, now with a certain resignation:

e) If a, b, c and d are valid, z is valid.[5]

Achilles cannot win this debate. If the tortoise chooses not to accept the first use of *modus ponens*, no single, additional statement will ever satisfy him. Achilles will be forced into an infinite regression of additional logical axioms.

This is not an idle point; indeed, the leading twentieth-century proponent of intuitionism, Brouwer, did not accept one commonly accepted principle of argumentation, proof by contradiction, because it was not constructive. Brouwer believed that a mathematical object should be accepted only if it could be constructed, or found, in a finite number of steps. To give but one example of a number Russell and Whitehead would accept and Brouwer would not, we use the decimal expansion for π. Since π is not a rational number, its expression as a decimal is unending and never settles into a repeating pattern—somewhere in its decimal expansion π may or may not contain the consecutive digits 123456789. We let k denote the number of decimal digits of π before this sequence of numbers appears. If it never appears we let k be 0. So k is either an even counting number, an odd counting number, or 0. Using k, we let $P = (-1)^k$. Since k is either even, odd, or 0, the formula for P tells us that it is either $+1$ or -1. Brouwer would not accept the existence of the number P, unless you could determine the value k. And if the sequence 123456789 happens never to appear in π, then k cannot be determined in a finite number of steps.

Nonetheless, for the rest of this chapter we will assume that everyone accepts the same deductive procedures, because the other difficulties with our operational definition of mathematical truth are more serious—they show that mathematical certainty cannot be obtained through the deductive method, even if everyone were to agree on a common list of axioms and rules of logic.

CONSISTENCY AND COMPLETENESS

The discovery/invention of non-Euclidean geometry led mathematicians to scrutinize Euclid's *Elements* even more closely than they had before. Over time, mathematicians realized that Euclid had employed some axioms without either stating them or acknowledging that he was not stating them. We give but one example: A line segment that crosses one side of a triangle, not at a vertex, must, when extended, cross another side of the triangle. When this assertion is illustrated (Figure 12.1) it seems to be obvious.

FIGURE 12.1. An example of a seemingly obvious geometric truth that cannot be deduced from Euclid's axioms: A line that intersects one side of a triangle, not at a vertex, must intersect the triangle at another point.

However, this result cannot be deduced from Euclid's postulates and definitions. The formalist mathematician Hilbert provided a new set of axioms for Euclidean geometry. In Hilbert's scheme (there have been others) Euclidean geometry rests on twenty-one axioms. These include the one above, as well as others concerning *betweenness* and *connectedness*.

Hilbert's understanding of the meaning of his axioms of geometry was not the same as Euclid's. For Euclid, axioms were self-evident truths about geometry that were based on our experience with physical space; for Hilbert, they were the beginning point for the deduction of new results that were logical consequences of the axioms. Hilbert did not claim his axioms necessarily had anything in common with our assumptions about the geometry of our world; they were simply the starting assumptions for one of an unlimited number of possible mathematical systems. This is the view Hilbert expressed when he said that mathematics "is nothing more than a game played according to certain simple rules with meaningless marks on paper."

But Hilbert should not be viewed as someone who thought mathematics was meaningless; he just felt that the only certainty we could have would come from formally deducing results from clearly stated axioms (without regard to the mathematical intuitions that led to the axioms in the first place). In Hilbert's view, the most important property for a list of axioms is not that it corresponds to some mathematical reality, whose existence he rejected, but that the results deduced from the axioms on the list should not contradict one another. This is the notion of the *consistency* of a set of axioms; it is the idea that there is not embedded deep within the list of axioms some contradictory assumptions that will eventually lead to contradictory theorems.

This is an important place where Hilbert differs from more Platonically inclined mathematicians. For Platonists, such as Russell and Whitehead, an axiom is true if, harking back to Aristotle's definition of truth, "it agrees with (the Platonic) reality." If all of the axioms are true, then any statements deduced from them will be true, and so the list of axioms will automatically be consistent. They believed that there could not be any contradictory results or relations in the Platonic realm. But Platonically inclined mathematicians do not have any way to check the validity of what are perceived to be self-evident truths, and so they too sought to establish the consistency of their axioms. If their axioms were consistent, they, at least possibly, could be true.

Consistency is not the only property Platonic mathematicians expect of their system of axioms. These axioms should also be diverse enough to allow for the eventual deduction of all truths about the realm of mathematics. This attribute is called the *completeness* of the list of axioms. Hilbert was not overly concerned with the completeness of his axiomization of Euclidean geometry since, for him, the true statements are those, and only those, that can be deduced from the axioms.

Completeness was the original goal of the logicist program. Employing the concept of set in a fundamental way, Russell and Whitehead sought to base mathematics on logic—they even sought to reduce mathematics to logic. Although Russell and Whitehead were both Platonists, they gave definitions for mathematical notions in terms of sets. It took them many pages to define the number "one." From these basic definitions, Russell and Whitehead hoped to remove any ambiguity from mathematics and thus obtain greater certainty.

KURT GÖDEL

The great tragedy of Science—the slaying of a beautiful hypothesis by an ugly fact.
— *Thomas Huxley, "Address to the British Association for the Advancement of Science" (1870)*

Long before mathematicians became concerned with the consistency or completeness of general axiomatic systems, writers had examined the analogues of these ideas in fiction. These examinations were performed not just by literary theorists but by the writers themselves. This exploration of the limits of fiction through fiction is called *metafiction*, and looking at two examples will help to elucidate what was at stake for mathematicians.

In order to illustrate the parallels between metafiction and Hilbert's and Russell's concerns with axiomatic systems, one has to take a fairly naive view of what narrative fiction is all about. In particular, we will have to make the questionable assumption that in a work of fiction there is a reality being described by the narration, or at least a reality the author wants to convey to the reader. We posit the following correspondences:

FICTION		MATHEMATICS
the text	↔	a collection of mathematical statements
statements from the text the reader believes are true	↔	axioms
conclusions of the reader about the reality underlying the text	↔	deduced mathematical results
the reality behind the text	↔	the Platonic realm of mathematics

Writers have explored the relationships between each of the four areas in the left-hand column, as well as their relationship with the real world, but the only ones that will concern us are the relationships between "statements from the text the reader believes are true" and the next two, "conclusions of the reader about the reality underlying the text" and "the reality behind the text."

If the *axioms* of the text are consistent, then a diligent reader should be able to draw true conclusions about the text. But in Coover's "The Babysitter" we saw that it would not be possible for the reader to know whether or not his conclusions are consistent (see chapter 8). The reader of "The Babysitter" cannot know which of the narrations in the story contradict one another, so the reader cannot even know what is happening with the characters.

The completeness, or rather incompleteness, of the axioms (of the text) has been portrayed in fiction almost since its inception. It is simply the observation that the reader cannot know everything about the story behind the narration, no matter how detailed the narration is. One technique writers have used to make the reader aware of this deficiency is called *frame-breaking*. This occurs when the author interrupts the narrative to make some comment. We have already seen one example of this. In the novel *Tristram Shandy* the protagonist not only tells his story but also discusses the process of writing his story (see chapter 6). A more modern example of frame-breaking occurs in the novel *The French Lieutenant's Woman*, by John Fowles. In order to emphasize that the reader can never know the entire truth, Fowles offers two endings to the story and he does not tell us which he intends the reader to believe. The film version of *The French Lieutenant's Woman* masterfully offers these alternate endings by telling two stories in parallel.

But mathematicians were not so willing to incorporate ambiguities into theorems, which, after all, represent truths. Hilbert attempted to prove the consistency of his list of axioms for Euclidean geometry. He could not. But in 1904 Hilbert used Descartes' coordinate geometry to demonstrate that if there is a consistent set of axioms for arithmetic then his axioms for Euclidean geometry are consistent. Late in the nineteenth century the mathematician Giuseppe Peano (1858–1932) had axiomatized arithmetic; however, neither Hilbert nor anyone else could demonstrate the consistency of Peano's axioms. This left mathematicians with the unsettling knowledge that someday, mathematicians might prove theorems in geometry, or arithmetic, that contradict one another.

Twenty-seven years after Hilbert had shown that the consistency of arithmetic would imply the consistency of Euclidean geometry, a young mathematician published a paper that forever undermined mathematical certainty. In his paper "On Formally Undecidable Propositions in *Principia Mathematica* and Related Systems," Kurt Gödel (1906–78) showed that both the Russell-Whitehead attempt to reduce mathematics to logic, in *Principia Mathematica* (1910), and Hilbert's attempt to axiomatize mathematics were doomed. Gödel showed that there are features inherent to axiomatic systems that limit their access to truth. (As Gödel was a Platonist, his results probably did not overly disturb him; his research had only revealed that the world of mathematics is too rich to be fully understood through axiomatic systems.)

One of Gödel's theorems, and it really is a theorem, was that the proposition "the axioms of arithmetic are consistent" can neither be proved nor disproved unless one works in a larger system. This is an example of what is referred to in the title of Gödel's paper as an "undecidable proposition." As a practical matter this means that results deduced from any list of axioms for arithmetic might not all be true (since two might contradict each other). Thus a mathematician is always in the position of a person reading a work of fiction, never knowing if the author has provided a consistent narrative.

Gödel's second theorem doomed Hilbert's program. What Gödel *proved*, if you believe, as Gödel did, that there is some Platonic realm of mathematical objects and relationships, is that there are true mathematical statements that will never be uncovered through the application of deductive reasoning within an axiomatic system. Two examples will illustrate how Gödel's theorems manifest themselves in mathematicians' search for mathematical truth.

The first example is based on Cantor's especially elegant conception of mathematical infinity. Cantor's use of completed infinite processes led to a meaningful way to discriminate between different sizes of infinite sets. Although both the set of all counting numbers and the set of all numbers in the continuum are infinite, Cantor's theory makes it possible to say that the number of numbers in the continuum, c, is greater than the number of counting numbers, $\aleph_0$. Symbolically,

$$\aleph_0 < c.$$

The set of all counting numbers and the set of all numbers on the number line are familiar to us, but we know of other infinite sets, such as the collections of all even counting numbers, the collection of all numbers whose decimal expansions involve only 0s and 1s, and the collection of all algebraic numbers. Cantor's method makes it possible to compare the cardinalities of each of these infinities with $\aleph_0$ and c, and this work leads to the surprising result that each of their cardinalities is either $\aleph_0$ or c. Indeed, if you look at any run-of-the-mill infinite set, its cardinality will be one of the two infinities $\aleph_0$ or c. Anyone working through many examples probably would reach the conclusion that there is no infinite set whose cardinality is greater than $\aleph_0$ yet less than c. This conclusion is now known as the *continuum hypothesis*.

Cantor attempted to prove this result but could not. Gödel also attempted to establish that there is no cardinal number between $\aleph_0$ and c, and like Cantor, Gödel could not find a proof. But Gödel then took another approach; he wanted to know whether or not the continuum hypothesis was a reasonable thing to even believe. What Gödel proved was that if the standard axioms of set theory are consistent, and if the continuum hypothesis is assumed to be true, then you will not be able to deduce a contradiction. In other words the continuum hypothesis is consistent with the other axioms. In the 1960s the mathematician Paul Cohen (1934–2007) supplemented Gödel's theorem with a theorem of his own: Cohen showed that if you assume the continuum hypothesis is *false*, you still will never be led to a contradiction. Putting these two results together, mathematicians were led to a disappointing conclusion: Gödel and Cohen had shown that mathematicians could not use their commonly accepted axioms to determine whether the continuum hypothesis is true or false. There can be two different types of mathematics—one where the continuum hypothesis is true and one where the continuum hypothesis is false.

Thus, Gödel's incompleteness theorem is not just a theoretical curiosity; it shows that there are simply stated mathematical questions whose answers cannot be provided within the framework of the commonly accepted axiomatic system. Modern mathematicians are in much the same position as the early Greeks—our assumptions do not allow us to fully understand infinite collections. Cantor's conception of mathematical infinitude answered some questions, such as why the notion that the whole is greater than the part does not extend beyond finite collections to infinite ones, but introduced new questions, such as, "is the continuum hypothesis true?"

Our second example also deals with the existence of mathematical entities associated with infinity—not with the infinitely large, but with the infinitely small indivisibles (infinitesimals), which were investigated by Abraham Robinson (1918–74) in the 1960s. To appreciate Robinson's work on infinitesimals, we need to understand the modern view of axiomatic systems, and the simplest example comes not from Euclidean geometry but from the axioms for the counting numbers as given by Peano. Peano's axioms are considered to be a minimal list of true statements about the integers, and if this appealingly short list is *complete* then all true statements about the integers can be derived from it. Although they should be stated in formal, mathematical symbols, Peano's five axioms are essentially the five assumptions

Axiom 1. 1 is a counting number.

Axiom 2. Every counting number n has a successor, usually denoted by $n + 1$.

Axiom 3. Every number, except 1, has a predecessor.

Axiom 4. Two different numbers cannot have the same successor.

Axiom 5. The axiom of mathematical induction: Suppose we are presented with a collection of counting numbers, which we will denote by S. If we can show that 1 is in S and that S contains the successor of each number in it, then S contains all of the counting numbers.

These axioms describe what we think of as the positive, counting numbers, they begin with 1, which has as its successor 2, which has the successor 3, and so on. Thus we obtain the sequence:

$$1 \to 2 \to 3 \to 4 \to 5 \to 6 \to 7 \to 8 \to 9 \to 10 \to \ldots$$

If we only assume the first two axioms we could have a repetitive list of successors:

$$1 \to 2 \to 3 \to 1 \to 2 \to 3 \to 1 \to 2 \to 3 \to 1 \to$$
$$2 \to 3 \to 1 \to 2 \to 3 \to 1 \to 2 \to 3 \to$$

which is a finite system consisting of the three symbols 1, 2, and 3. Note that in this system 3 is a predecessor of 1, symbolically $3 + 1 = 1$, which the third axiom prohibits.

However, even assuming each of the first three axioms, it would still be possible to have a repeating list of successors

$$1 \to 2 \to 3 \to 4 \to 2 \to 3 \to 4 \to 2 \to 3 \to$$
$$4 \to 2 \to 3 \to 4 \to 2 \to 3 \to 4 \to$$

and so a finite system. This system consists of the symbols 1, 2, 3, and 4. The symbol 2 has two predecessors, 1 and 4, something that is prohibited by the fourth axiom.

The last axiom seems to tell us that, as in the Pythagorean number system, 1 is the generator of all counting numbers; in other words that there cannot be a counting number w not contained in the list:

$$1 \to 2 \to 3 \to 4 \to 5 \to 6 \to 7 \to 8 \to 9 \to 10 \to \dots$$

But it does not. There is nothing in these five axioms to prohibit the existence of, for example, a *counting number* w that is larger than all the counting numbers that can trace their origin back to 1. In other words, there is nothing to exclude the existence of an infinitely large element.

Thus Peano's axioms allow for what are called *nonstandard models*, which means they allow for mathematical systems that do not match our intuition for what the standard counting numbers are. These nonstandard models are important for the discovery of new truths about the counting numbers, since the counting numbers are a small part of any of them. So, working within this system, a mathematician might be able to use the additional elements in the model to prove things about the counting numbers.

It is not too difficult to understand how Abraham Robinson established a model for the axioms of the real numbers that allows for infinitesimals, once we see how the real numbers are obtained from the counting numbers. For simplicity, we only discuss positive numbers. The positive rational numbers (the positive fractions) are obtained by taking all ratios of the counting numbers; the positive rational numbers consist of all numbers of the form a/b, where a and b are positive counting numbers. To obtain the positive real numbers, we employ the idea of convergence from chapter 11; a real number is obtained in much the same way as a mathematical fractal is obtained—a real number is defined to be the thing you get when you look at a convergent sequence of rational numbers. Using this process, if we start with the standard model for the counting numbers, 1, 2, 3, . . . we end up with what we think of as the usual positive real numbers—numbers that correspond to the lengths of geometric magnitudes. However, if we start with a nonstandard model for the counting numbers, for example one that contains a number w that is larger than any standard (ordinary) counting number, we obtain the positive rational number $1/w$. This rational number satisfies the property

For any "standard" counting number n, $1/w < 1/n$.

It can be deduced from the axioms that $1/w$ is a positive quantity that is smaller than the length of any geometric magnitude. So $1/w$ is truly an infinitesimal. Then, defining the real numbers to be all of the entities obtained from convergent sequences of nonstandard rational numbers, we obtain a system of numbers with an unlimited number of quantities between any two "standard" numbers. In particular, we obtain an unlimited number of infinitesimally small quantities.

If these infinitesimal quantities are interpreted geometrically, then Robinson has rediscovered what Nicolas de Cusa, Kepler, and Leibniz believed: that a line segment or curve consists of infinitely many infinitely short segments. Kepler had used these to establish geometric theorems, such as the formula that the area of a circle equals one-half of its radius times its circumference, and Leibniz had used infinitesimals to develop his version of calculus.

Even though Robinson showed that the existence of infinitesimals does not contradict any of the generally accepted axioms of mathematics, few working mathematicians accept them as being "real" geometric magnitudes.

These two examples illustrate what Nelson was referring to when he wrote that no real progress has been made on the question of ontology. For example, mathematicians have no guidance on whether they should accept the existence of infinitesimals, and as Gödel and Cohen showed, they will never have any guidance on whether there can be an infinite set whose cardinality is wedged between $\aleph_0$ and c. By appealing to plenitude, mathematicians embrace most objects they uncover. However, infinitesimals have not entirely taken hold, possibly because they defy intuition—they seem to allow for the existence of geometric magnitudes shorter than any rational length. But mathematicians do not dwell on this. Nelson went so far as to conclude his article with, "In mathematics, the ontological quest is misconceived and should be abandoned."[6]

If the ontological quest is to be abandoned, why are there so many mathematicians sitting in their offices working so hard? Of course, that question has a different answer for each mathematician, and it would have a great deal to do with their individual histories, but a safe generalization is that they mostly work so hard not because they are concerned with ontology but because they are concerned with truth.

TRUTH

The criterion of the modern artist is Truth rather than Beauty, and to this extent modern art is still keeping pace with natural science.
— Herbert Read, "Human Art and Inhuman Nature" (1955)

If logical analysis of mathematical concepts and the relationships between them can never reveal all mathematical truths, how will they ever be known? Before considering this question, it is important to describe what a working mathematician does—a mathematician seeks not only to prove mathematical truths but also to develop an understanding of some part of the mathematical world. This understanding almost never comes from a *proof*; it comes from examining examples and looking for patterns.

Suppose you are a mathematician and you become fascinated with right triangles, whose sides all have lengths that are whole numbers, such as the three triangles in Figure 12.2.

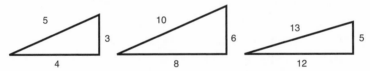

FIGURE 12.2. Three right triangles whose sides are all whole numbers.

You work very hard and discover more and more of these triangles. There seems to be no limit to how many such triangles you can produce. Some of them have the same shapes, such as the three-four-five triangle and six-eight-ten triangles above, but among the unlimited number of such triangles there also seem to be an unlimited number of different shapes. Being a mathematician, you naturally ask yourself the question any mathematician would ask: How many triangles are there whose sides are all whole numbers?

Having posed this question, you stop your work for the day and prepare your morning classes. The next day at lunch you happen to mention your question to another colleague who answers, "There are an unlimited number of whole-number-sided right triangles, and there are even formulas for the lengths of their sides." Your colleague writes the formulas on a napkin and hands it to you. You are impressed, and maybe a bit deflated, but you thank your colleague and continue with your day. However, you are not satisfied with your colleague's formulas, because your original question was not only how many triangles are there whose sides are all whole numbers, but how many triangles are there whose sides are all whole numbers and why? It is the *why* your colleague's formulas fail to address, and this is the part of the question that separates an artist from a craftsman and a mathematician from a computer.

In an attempt to answer the *why* portion of your question, you return to your desk and continue experimenting. Drawing more triangles is not going to help; you are already overwhelmed by the range of numbers appearing as sides to right triangles, so you need to try something else. This is where the mathematician becomes a creative artist; you need a new idea, and you need to find another way to think about right triangles. You decide to try something new—you decide to put all of the triangles you have found into forms that will allow you to compare them. In particular, you know that each triangle in Figure 12.2 is similar to a triangle whose hypotenuse equals one unit, so you write down all of these triangles. The original triangles that had the same shape reduce to the same triangle, but as before, you find that there are an unlimited number of different shapes:

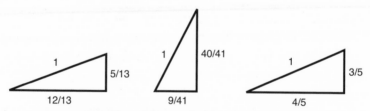

FIGURE 12.3. Three right triangles whose hypotenuse equals one unit and whose two legs are rational numbers.

Still in an experimental mode, you superimpose these triangles over each other:

FIGURE 12.4. Right triangles, as in Figure 12.3, inside a circle whose radius equals one unit.

The top vertices of these triangles seem to form an arc and you realize that they all lie on a circle whose radius is one unit. In an instant you verify that they do indeed all lie on a circle. (It is a consequence of the Pythagorean theorem.) You have uncovered a deep connection: Each right triangle, with integral sides, corresponds to a point whose coordinates are fractions that lie on the circle whose radius equals one. Moreover, each point on the circle whose coordinates are fractions yields a right triangle whose sides have integral lengths. The reason there are an unlimited number of integral right triangles is because of properties of the circle—you already know from some other investigation that there are an unlimited number of points on the circle whose coordinates are rational numbers (and the formulas for finding these points are exactly the ones your colleague had handed you earlier).

A mathematician would call this connection between integral-sided right triangles and points on the circle with rational coordinates beautiful. And mathematicians associate beauty not only with formulas or even with geometry—any mathematical relationship can be considered to be beautiful. In the above example, the simplicity and definitiveness of the connection between the triangles and the points on the circle make it beautiful. It is such a surprising connection that even without having any formulas for points on the circle, you knew the moment you uncovered it that you had found some deeper, hidden truth. An aesthetic adjective has emerged as the one mathematicians attach to such beauty; it is elegance.

ELEGANCE

In his book *The Elegant Universe*, Brian Greene wrote: "[Einstein felt] that general relativity describes gravity with such a deep inner elegance, with such simple yet powerful ideas, that he found it hard to imagine that nature could pass it by. General relativity, in Einstein's view, was almost too beautiful to be wrong."[7] Just as physical ideas can be elegant, with the "simple yet powerful ideas" of general relativity being but one of its manifestations, mathematics has an aesthetic of elegance of its own. Mathematical elegance takes several forms, and mathematicians do not always agree on the aesthetic appeal of a particular discovery or conjecture. But three types of relationships are almost universally held to be elegant:

1. those that are especially simple yet definitive, such as Euler's formula relating the number of vertices, edges, and regions of a doodle (see chapter 8),
2. those that provide a framework within which the previously incomprehensible becomes comprehensible, such as Cantor's theory of mathematical infinity based on the simple notion of a one-to-one correspondence (see chapter 11), and
3. those that reveal hidden connections between previously unrelated objects, such as the relationship between right triangles and points on a circle that leads to a description of all integral right triangles.

The elegance of a result does not depend on having its proof—its elegance is intrinsic to it, and the elegance of a result implies that the result must be true. Many mathematicians will disagree with this last sentence, but in the past thirty years major mathematical prizes have been awarded to mathematicians, at least in part, for their *insights*—another code word for having discovered an elegant mathematical relationship. Two of these discoveries involved the elucidation of a previously unknown, or unappreciated, relationship between very abstract mathematical objects—these are the discoveries of Robert Langlands in the 1960s and Paul Vojta in the 1980s. Langlands and Vojta both proceeded like our industrious, right-triangle-obsessed mathematician. They worked out examples, proved theorems that became *special* cases, and then in a moment of revelation perceived that there might be a more general relationship in the background. By working out even more examples and deducing more results, they gained more insight. The details of the discoveries of these mathematicians are too advanced for us to discuss them here, but they have both been used to uncover new truths in number theory by exploiting their proposed correspondences.

A third insight was the proposal that there exists a unified framework within which to mathematically examine both the geometry of space and quantum mechanics. This was Edward Witten's suggestion in the 1990s: that the different versions of string theory that had been proposed could all be taken to be manifestations of a single underlying theory, his *M-theory*. Physicists were overwhelmed by the simplicity of Witten's vision. The physicist John Schwarz is quoted in Greene's book as saying that the "mathematical structure of string theory was so beautiful and had so many miraculous properties that it had to be pointing toward something deep."[8] Physicists hope that it might someday be possible to test Witten's vision experimentally, to demonstrate that its mathematical beauty is a manifestation of physical truth. But mathematicians have no physical reality against which to test the insights of Langlands and Vojta; they have already been tested and each of them is so definitive that its truth is assumed.

PROOF AND TRUTH

Since the gold standard for the widespread acceptance of a mathematical result is a proof, a final example illustrates that the conception of what a mathematical proof is is subtler than most mathematicians are willing to admit.

Each of the earlier examples dealt with ontology; our last example concerns itself more with the question of what is a proof. In the 1970s two mathematicians, Kenneth Appel and Wolfgang Haken, announced they had established the *four-color theorem*, which had withstood the efforts of many talented mathematicians since it had first been formulated a century earlier. The four-color theorem is simple to state: Take a map of countries or regions where each country or region consists of only one piece (unlike, for example, Michigan) and color the map so that countries or regions having a common border are different colors. (If two regions only meet at a point, such as Arizona and Colorado, they can have the same color.) The four-color theorem states that it is possible to color any such map using four or fewer colors. The simple "map" in Figure 12.5 requires four colors, since each region touches all of the other regions, and so illustrates that sometimes four colors are needed.

FIGURE 12.5. This "map" of four countries demonstrates that some maps require four colors to color all of the regions so that no two adjacent regions are the same color.

In their attempts to prove the four-color theorem, mathematicians had discovered that there is a small set of configurations of countries, at least one of which must be present in any map (these are known as *unavoidable configurations*). For example, any map you look at will contain at least one country with five or fewer neighbors. The proof of this type of a result is not unlike that of Euler's formula for doodles (see chapter 8). Mathematicians came to understand that if each of these unavoidable configurations could be colored with four colors, then so could any map, so the hunt was on to find them all. (This is not entirely accurate as all mathematicians needed to find were the *reducible unavoidable configurations*; for a somewhat nontechnical description of the history of this search the reader should consult Robin Wilson's book *Four Colors Suffice*.)[9] This is precisely what Appel and Haken, did in 1976, but their proof used a computer in an essential way. Not only did the computer check that these reducible unavoidable configurations can be colored with four colors, but the computer was programmed to find the configurations.

The proof of the four-color theorem does not show mathematicians *why* the theorem is true; it does not offer any insights into the nature of maps. Rather it is simply a checking of cases. This inelegance alone would have left many mathematicians with the feeling that we just do not understand the four-color theorem. But here the situation is even worse; no mathematicians will ever be able to study the proof to gain any insights. The answer to the question of why the four-color theorem is true can only be answered with "because."

What are needed to resolve the continuum hypothesis, to make infinitesimals more acceptable, or to increase the comfort level of mathematicians with the proof of the four-color theorem are new insights, à la Langlands and Vojta. Only an elegant idea, whether it concerns the nature of infinite sets or of maps drawn on a piece of paper, can change the present impasse. Mathematicians have probably gone as far as they can with their present assumptions, both implicit and explicit. They are a bit like the Italian algebraists who acknowledged, but did not accept, *imaginary* solutions to algebraic equations. However, it is an article of faith that no portion of the Platonic realm of mathematics can forever remain mysterious; we only have to await the announcement by some solitary researcher of a new insight into elusive mathematical truth.

NOTES

Chapter 1. Mysticism, Number, and Geometry

1. Gafori, 1451–1522.

2. Koestler, *The Sleepwalkers*, 25.

3. Macrobius, *Commentary*, 186.

4. Boethius, *De musica*.

5. Sidney, *Defense of Poesy*, 58.

6. Shakespeare, *Merchant of Venice*, 5.1, ll. 58–62.

7. Donne, "Hymn to God, My God, in My Sickness" (1635), ll. 1–5.

8. Heninger, *Touches of Sweet Harmony*, 388.

9. Langer, *Feeling and Form*, 27.

10. Ginsberg, "Lysergic Acid," in *Kaddish*, 86, ll. 10–12.

11. Ovid, *Metamorphoses*, bk. 15, ll. 60–65.

12. Iamblichus, "Protrepticae Orationes ad Philosophiam," 411.

13. Dacier, *Life of Pythagoras*, xiii.

14. Plutarch, *Plutarch's Morals*, 1:17, p. 28.

15. Diogenes Laertius, *Lives, Opinions, and Remarkable Sayings*, vol. 2, bk. 8, p. 23.

16. Plutarch, *Plutarch's Morals*, 1:17, p. 29.

17. Cicero, *On Divination*, 171.

18. Empedocles, *Poem*, 265.

19. Plato, *Timaeus*, 58.

20. Ovid, *Metamorphoses*, bk. 15, ll. 163–67.

21. Salluste du Bartas, *His Devine Weekes and Workes*, "The Columns: The Fourth Part of the Second Day of the Second Weeke," 155, ll. 104–9.

22. Ibid., ll. 118–21.

23. Ibid., l. 136.

24. Aristotle, *Metaphysics*, bk. 1, pt. 5.

25. Simplicius, *On Aristotle's* On the Heavens, 51

26. Theon of Smyra, "Expositio rerum mathematicarum ad legendum platonem utilium," quoted in Cornford, *Plato's Cosmology*, 70. The connection between these four "simple bodies" and the four elements—earth, air, fire, and water—is examined in chapter 2. In Plato's *Republic*, book 4, the three parts of the soul are said to be *logos* (mind and conscious awareness), *thymos* (emotion), and *pathos* (appetite).

27. Pound, "Canto XCI," in *Section*.

28. Creeley, "Four," ll. 11–12.

29. Augustine, *On Free Choice of the Will*, 53.

30. Galileo Galilei, *Dialogue*, 11.

31. Ibid., 12.

32. Agrippa von Nettesheim, *Three Books of Occult Philosophy*, 310.

33. Chaucer, "The Knight's Tale," *The Canterbury Tales*, pt. 3, ll. 1182–87.

34. Kepler, *The Secret of the Universe*, 149.

Chapter 2. The Elgin Marbles and Plato's Geometric Chemistry

1. Lovejoy, *The Great Chain of Being*, 25.

2. Heisenberg, *Philosophic Problems of Nuclear Science*, 59.

3. Letter dated November 22, 1817, in Keats, *Letters*, 1:184.

4. Keats, "Ode on a Grecian Urn," ll. 46–50.

5. Kepler, *The Secrets of the Universe*, 149–51.

6. Livio, *The Golden Ratio*, chapter 7.

7. Plato, *Timaeus*, 48.

8. Augustine, *Concerning the City of God*, bk. 8, pp. 308, 312.

9. Augustine, *Letters*, 4:19.

10. Spenser, "An Hymn in Honour of Love," ll. 78–84.

11. Heraclitus, *Fragments*, 15.

12. Empedocles, *Poem*, 231.

13. Taylor, *The Atomists*, 9.

14. Ibid.

15. Ibid., 13.

16. Plato, *Timaeus*, 44.

17. Ibid.

18. Ibid.

19. Heraclitus, *Fragments*, 17.

20. Plato, *Timaeus*, 68.

21. Ibid.

22. Ibid., 78–79.

23. Ibid., 78.

24. Ibid., 80.

25. According to Plato, "When earth meets fire it will be dissolved by its [fire's] sharpness, and whether dissolution takes place in fire itself or in a mass of air or water, will drift about until its parts meet, fit together and become earth again; for they can never be transformed into another figure" (ibid., 79–80).

26. Aristotle, *Physics*, bk. 1.

27. Plato, *Timaeus*, 44.

28. Macrobius, *Commentary*, 105.

29. Ibid.

30. Milton, *Paradise Lost*, bk. 2, ll. 890–900.

31. Sidney, *Defense of Poesy*, 6.

32. Ibid., 7.

33. Jonson, *Works*, 7:147.

34. Jonas, *The Divine Science*, 17.

35. Quoted in Jonas, *The Divine Science*, 4.

36. Jonas, *The Divine Science*, 4–5.

37. Sidney, *Defense of Poesy*, 11.

Chapter 3. An Introduction to Infinity

1. Vincent van Gogh to Theo van Gogh, Aug. 6, 1888, www.vggallery.com/letters/635_V-T_518.pdf.

2. First published in *Gilchrist's Life of William Blake*, 1863. It was edited from a manuscript written by Blake probably during his stay at Felpham (1800–1803). Blake did not restrict his attention to metaphysical infinity; in *The Four Zoas* (c. 1795–1804) he poetically appealed to the mathematical infinity: "She also took an atom of space & opened its center / Into Infinitude & ornamented it with wondrous art" (1:270).

3. Tennyson, "Flower in the Crannied Wall" (1869).

4. Neruda, "Tonight I Can Write," ll. 25–26.

5. Parmenides, "On Nature," quoted in Most, "Poetics of Early Greek Philosophy," 354, ll. 22–33.

6. Parmenides, "On Nature," quoted in Sedley, "Parmenides and Melissus," 121, ll. 44–49.

7. Ibid., 126.

8. Aristotle, *Physics*, bk. 3, sec. 5, p. 66.

9. Quoted in Lovejoy, *The Great Chain of Being*, 62.

10. Ibid.

11. Nicolson, *Mountain Gloom and Mountain Glory*, 135. The full list of attributes is ut Unum, Simplex, Immobile, Aeternum, Completum, Independens, A se existens, Per se subsistens, Incorruptibile, Necessarium, Immensum, Increatum, Incircumscriptum, Incomprehensible, Omnipraesens, Incorporeum, Omnia permeans et complectens, Ens per Essentiam, Ens actu, Purus Actus [as one, simple, unchanging, eternal, perfect, independent, existing by itself, subsisting by means of itself, incorruptible, necessary, immeasurable, uncreated, unbounded, incomprehensible, ubiquitous, bodiless, permeating and surrounding everything, essential being, being in actuality, pure actuality], translated by E. Christian Kopff, University of Colorado, Boulder.

12. *Spectator* 412 (Monday June 23, 1712).

13. Kant, *Critique of Judgment*, 2:26, 98.

14. Ibid., 95.

15. Aristotle, *Physics*, bk. 3, sec. 8, p. 76.

16. Lucretius, *The Nature of the Universe*, 56.

17. Borges, "The Library of Babel," in *Labyrinths*, 58.

18. Escher, *Escher on Escher*, 124.

19. Aristotle, *Physics*, bk. 3, sec. 6, p. 72.

20. Ibid., bk. 6, sec. 2, p. 141.

21. Escher, *Escher on Escher*, 41.

22. Aristotle, *Physics*, bk. 8, sec. 8, p. 219.

23. Ibid., bk. 6, sec. 9, p. 161.

24. Barth, "The Literature of Exhaustion," 74.

25. Aristotle, *Physics*, bk. 6, sec. 9, p. 161.

26. Ibid., 161–62.

27. Bergson, *An Introduction to Metaphysics*, 1.

28. B. Russell, "Recent Work on the Principles of Mathematics," 91.

29. Boccioni, "Plastic Dynamism," 93.

Chapter 4. The Flat Earth and the Spherical Sky

1. Aristotle, *On the Heavens*, 2:14.

2. Ibid.

3. Ibid., 2:13.

4. Cosmas Indicopleustes, *The Christian Topography*.

5. Ibid., 133.

6. Ibid., 244.

7. Copernicus, *On the Revolutions*, 3–5.

8. Augustine, *Concerning the City of God*, 664.

9. Ibid.

10. Engels estimates that 1 stade = 606 ft., 10 inches. Engels, "The Length of Eratosthenes' Stade," 298–311.

11. Aristotle, *Physics*, bk. 4, sec. 2.

12. Plato, *Timaeus*, 46.

13. Aristotle, *Physics*, bk. 8, sec. 8, p. 214.

14. Plato, *The Republic*, 340.

Chapter 5. Theology, Logic, and Questions about Angels

1. Boethius, "De hebdomadibus," 299.

2. Ibid., 300. Although an axiom is supposed to be a self-evident truth, Boethius offered a justification for his second axiom, "For being itself does not exist yet, but

that which is exists and is established when it has taken on the form of being" ("De hebdomadibus," 300).

3. Anselm, *Proslogium*, 2:8.

4. Robinson, *Readings in European History*, 1:450–51.

5. Ibid.

6. Averroes, 1126–98.

7. Grant, *God and Reason*, 192; Albertus quoted in Synan, "Introduction: Albertus Magnus," 9.

8. Quoted in Lohr, "Medieval Interpretation of Aristotle," 90.

9. Thorndike, *University Records*, 64–65.

10. Grant, *Science and Religion*, 181–82.

11. Carnes, *Axiomatics and Dogmatics*.

12. Collins, "Questions about Angels," from *Questions about Angels*, 25–26, ll. 24–26.

13. Dante, *The Divine Comedy*, *Paradise*, canto 1, ll. 100–110.

14. Dante, *Inferno*, canto 4, ll. 129–35. This passage continues with "Zeno, and Dioscorides well read / In nature's secret lore. Orpheus I marked / And Linus, Tully and moral Seneca, / Euclid and Ptolemy, Hippocrates, / Galenus, Avicen, and him who made / That commentary vast, Averroes" (ll. 136–41).

15. Dante, *Paradise*, canto 2, ll. 50–51.

16. Plutarch, *Plutarch's Morals*, 3:157.

17. Grant, *Planets, Stars, and Orbs*, 460–61.

18. Quoted in Duhem, *Medieval Cosmology*, 481.

19. Dante, *Paradise*, canto 2, ll. 73–82.

20. Grant, *Planets, Stars, and Orbs*, 527.

21. Dante, *Paradise*, canto 2, ll. 133–48.

22. Ibid., canto 8, ll. 29–31.

23. Ibid., ll. 39–44.

24. Van Helden, *Measuring the Universe*, 27.

25. Maimonides, *Guide for the Perplexed*, pt. 3, chap. 14, p. 277.

26. Carter, "Considering the Void," in *Always a Reckoning*, 121.

27. Buber, *Between Man and Man*, 136.

28. Duhem, *Medieval Cosmology*, 74.

29. Aquinas, quoted in Duhem, *Medieval Cosmology*, 12.

30. Lovejoy, *The Great Chain of Being*, 52.

31. Ibid.

32. Aristotle, quoted in Lovejoy, *The Great Chain of Being*, 56.

33. Lovejoy, *The Great Chain of Being*, 59.

34. Bruno, "On the Infinite Universes and Worlds," 257.

Chapter 6. Time, Infinity, and Incommensurability

1. Plato, *Timaeus*, 54.

2. Aveni, *Conversing with the Planets*.

3. Grant, *Planets, Stars, and Orbs*, 499.

4. Ibid.

5. Three centuries later Kepler appealed to incommensurability in the motions of the heavens to demonstrate that these motions were not the work of God but the result of natural processes.

6. Peacock, "The Four Ages of Poetry," 324–33.

7. Spengler, *Perspectives of World-History*, 507.

8. William Butler Yeats, "The Second Coming" (1919), ll. 3–4.

9. Davidson, *Proofs*.

10. Ibid., 54–55.

11. Moore, *The Infinite*, 48.

12. White, *History of the Warfare of Science with Theology in Christendom*, 9.

13. Nicolson, *Mountain Gloom and Mountain Glory*.

14. Milton, *Paradise Lost*, bk. 7, ll. 282–87.

15. Salluste du Bartas, *His Devine Weekes and Workes*, "The Third Day of the First Weeke," ll. 34, 38–41.

16. Marvell, "Upon the Hill and Grove at Billborow," (c. 1651), ll. 9–12.

17. Donne, "An Anatomy of the World: The First Anniversary," (1611), ll. 300–301.

18. Moore, *The Infinite*, 44.

19. B. Russell, "Recent Work on the Principles of Mathematics," 96.

20. Sterne, *The Life and Opinions of Tristram Shandy, Gentlemen* in *Works of Laurence Sterne*, 1:311.

Chapter 7. Medieval Theories of Vision and the Discovery of Space

1. Vitruvius, *Ten Books on Architecture*, bk. 4, chap. 6.

2. Veltman, *Linear Perspective*, 34.

3. Andrews, *Story and Space in Renaissance Art*, 1.

4. Ibid., 5.

5. Quoted in Andrews, *Story and Space in Renaissance Art*, 35.

6. Piero della Francesca, *De prospectiva pingendi*.

7. An illustrated analysis of this painting is carried out by Loran, in *Cézanne's Composition*, plate 14.

Chapter 8. The Shape of Space and the Fourth Dimension

1. Marvell, "The Definition of Love," ll. 1–4, 21–28.

2. Escher, *Escher on Escher*, 42.

3. Grant, *Science and Religion*, 174.

4. Misner and Wheeler, "Classical Physics as Geometry," 526.

5. Antliff and Leighton, *Cubism and Culture*.

6. Stein, "Picasso."

7. Weber, "The Fourth Dimension from a Plastic Point of View," 25.

8. Duchamp, "A l'infinitif," in *Salt Seller*, 92.

9. Arensberg, "Arithmetical Progression of the Verb 'To Be,'" 3.

10. Weber, "The Fourth Dimension from a Plastic Point of View," 25.

11. Malevich, *The Non-Objective World*, 67, 68, 76.

Chapter 9. What Is a Number?

1. *De Stijl*, vol. 2, no. 1 (November 1918), 4.

2. Quoted in Joosten, "Painting and Sculpture in the Context of De Stijl," 60.

3. Van Doesburg, "Painting: From Composition to Counter-Composition," 205.

4. Vantongerloo, "Reflections III," in *Paintings, Sculptures, Reflections*, 20.

5. Ibid.

6. Ibid., 27.

7. Vantongerloo, "Introductory Reflections," in *Paintings, Sculptures, Reflections*, 3.

8. Cardano, *The Great Art*, 219–20.

9. Albert Girard, 1595–1632.

10. Quoted in Kline, *Mathematical Thought*, 1:253.

11. The persistence of this Pythagorean notion can be seen in a remark from Vantongerloo: "The number ONE is the only number having a value. But this one contains everything" ("Reflections I," in *Paintings, Sculptures, Reflections*, 10).

12. Hobson, *Squaring the Circle*, 4.

13. Donne, "Upon the Translation of the Psalms," ll. 1–4.

14. Bertuol, "The Square Circle of Margaret Cavendish," 24.

15. Ibid.

16. Cavendish, "The Circle of the Brain Cannot Be Squared," in *Poems and Fancies*, ll. 1–4.

Chapter 10. The Dual Nature of Points and Lines

1. Chevreul, *The Laws of Contrast of Colour*.

2. Aristotle, *Physics*, bk. 6, sec. 1, p. 138.

3. Ibid., sec. 4, p. 198. According to Grant one of the questions Scholastics attempted to answer was whether an angel "could be moved from place to place without passing through the middle [point]" (*God and Reason*, 277–78).

4. Augustine, *Concerning the City of God*, bk. 12, chap. 19, pp. 496–97.

5. Eldredge, "Late Medieval Discussions," 90–115.

6. Quoted in Duhem, *Medieval Cosmology*, 12.

7. Quoted in Murdoch, "Infinity and Continuity," 577.

8. Harclay quoted in Sylla, "God, Indivisibles, and Logic," 74.

9. Peter of Spain quoted in Duhem, *Medieval Cosmology*, 49–50.

10. Gregory quoted in Duhem, *Medieval Cosmology*, 52, 110–11.

Chapter 11. Modern Mathematical Infinity

1. B. Russell, "Recent Work on the Principles of Mathematics," 86.

2. Peitgen, Jurgens, and Saupe, *Fractals for the Classroom*, 182.

3. Taylor, Micolich, and Jonas, "Fractal Expressionism."

4. Ibid.

5. Cantor, *Contributions to the Founding of the Theory of Transfinite Numbers*, 85.

Chapter 12. Elegance and Truth

1. Quine, "On What There Is," 1.

2. Ibid., 13. Euclid's *Elements* contains a proof of the result that there exist infinitely many prime numbers, but the largest known prime number at that time could have been $2^{13} - 1 = 1,456$. In the seventeenth century Cataldi proved that $2^{17} - 1 = 131,071$ is a prime number, and in 1722, Euler showed that $2^{31} - 1$, which is greater than 2 billion, is a prime number.

3. Quoted in Eves, *Introduction to the History of Mathematics*, 571.

4. Bell, *Mathematics*, 21.

5. Borges, "Avatars of the Tortoise," in *Labyrinths*, 206–7.

6. Nelson, "Syntax and Semantics," 6.

7. Greene, *The Elegant Universe*, 166.

8. Ibid., 137.

9. Wilson, *Four Colors Suffice*.

BIBLIOGRAPHY

Abbott, Edwin. *Flatland: A Romance of Many Dimensions*. 6th ed. New York: Dover, 1952.

Agrippa von Nettesheim, Heinrich Cornelius. *Three Books of Occult Philosophy*. Translated by John Freake. London, 1651.

Albergotti, J. Clifton. *Mighty Is the Charm: Lectures on Science, Literature, and the Arts*. Washington, D.C.: University Press of America, 1982.

Andrews, Lew. *Story and Space in Renaissance Art: The Rebirth of Continuous Narrative*. Cambridge: Cambridge University Press, 1998.

Anselm. *Proslogium; Monologium; An Appendix in Behalf of the Fool by Gaunilon; and Cur Deus Homo*. Translated by Sidney Norton Deane. Chicago: Open Court, 1903.

Antliff, Mark, and Patricia Leighton. *Cubism and Culture*. London: Thames and Hudson, 2001.

Appel, K., and W. Haken. "Every Planar Map Is Four Colorable. I. Discharging." *Illinois Journal of Mathematics* 21, no. 3 (1977): 429–90.

Archimedes. "The Sand Reckoner." In *The Works of Archimedes*. Edited by T. L. Heath. Cambridge: Cambridge University Press, 1897.

Arensberg, Walter. "Arithmetical Progression of the Verb 'To Be.'" *391*, no. 5 (June 1917): 4.

Aristotle. *Categories and De Interpretatione*. Translated by J. L. Ackrill. Oxford: Clarendon Press, 1965.

———. *On the Heavens*. Translated by J. L. Stocks. Oxford: Clarendon Press, 1922. Web edition published by ebooks@Adelaide. http://ebooks.adelaide.edu.au/a/aristotle/heavens/.

———. *Metaphysics*. 2 vols. Translated by W. D. Ross. Oxford: Oxford University Press, 1924. Web edition published by ebooks@Adelaide. http://ebooks.adelaide.edu.au/a/aristotle/metaphysics/.

———. *Physics*. Translated by Robin Waterford. Oxford: Oxford University Press, 1996.

Augustine. *Concerning the City of God against the Pagans*. Translated by Henry Bettenson, with an introduction by David Knowles. Baltimore: Penguin Books, 1972.

———. *Letters*. Translated by Sister Wilfrid Parsons. 5 vols. New York, 1951–56.

———. *On Free Choice of the Will*. Translated by Thomas Williams. Indianapolis: Hackett, 1993.

Aveni, Anthony. *Conversing with the Planets: How Science and Myth Invented the Cosmos*. New York: Kodansha International, 1992.

Bacon, Roger. *The Opus Major of Roger Bacon*. Edited and translated by R. B. Burke. Philadelphia: University of Pennsylvania Press, 1928.

Baker, John Tull. *An Historical and Critical Examination of English Space and Time Theories*. 1670. Reprint, Bronxville, N.Y.: Sarah Lawrence College, 1932.

Barth, John. "The Literature of Exhaustion." In *The Friday Book*. New York: Putnam and Sons, 1984.

Bede. *Opera*. Basel, 1563.

Bell, E. T. *Mathematics: Queen and Servant of the Sciences*. London: G. Bell and Sons, 1952.

Benacerraf, P., and H. Putnam, eds. *Philosophy of Mathematics: Selected Readings*. Englewood Cliffs, N.J.: Prentice-Hall, 1964.

Bergson, Henri. *An Introduction to Metaphysics*. Translated by T. E. Hulme. New York: G. P. Putnam's Sons, 1912.

Bertuol, Roberto. "The Square Circle of Margaret Cavendish: The Seventeenth-Century Conceptualizations of Mind by Means of Mathematics." *Language and Literature* 10, no.1 (2001): 21–39.

Bierce, Ambrose. "An Occurrence at Owl Creek Bridge." In *The Norton Anthology of Short Fiction*. 2nd ed. Edited by R. V. Cassill. New York: W. W. Norton, 1981.

Bill, Max. *Georges Vantongerloo*. Brussels: Loiseau, 1980.

Boccioni, Umberto. "Plastic Dynamism, 1913." In *Futurist Manifestos*. Edited by Umbro Apollonio. New York: Viking Press, 1970.

Boethius. "De hebdomadibus." In Scott MacDonald, *Being and Goodness: The Concept of Good in Metaphysics and Philosophical Theology*. Ithaca: Cornell University Press, 1991.

———. *De Musica*. In *Patrologia Latina*. Vol. 63. Paris, 1882.

Borges, Jorge. *Labyrinths: Selected Stories and Other Writings*. New York: New Directions, 1964.

Browne, Thomas. *Sir Thomas Browne's Hydriotaphia and the Garden of Cyrus*. Edited by W. A. Greenhill. New York: Macmillan, 1896.

Bruno, Giordano. "On the Infinite Universe and Worlds." In *Giordano Bruno: His Life and Thought*. Translated by Dorothea Singer. New York: Schuman, 1950.

Buber, Martin. *Between Man and Man*. Translated by Ronald Gregor Smith. Boston: Beacon Press, 1959.

Bunim, Miriam S. *Space in Medieval Painting and the Forerunners of Perspective*. New York: Columbia University Press, 1940.

Butler, Christopher. *Number Symbolism*. New York: Barnes and Noble, 1970.

Cantor, Georg. *Contributions to the Founding of the Theory of Transfinite Numbers*. Translated by P. E. B. Jourdain. New York: Dover, n.d.

Cardano, Girolamo. *The Great Art; or, The Rules of Algebra*. Translated and edited by T. Richard Witmer. Cambridge: MIT Press, 1968.

Carnes, John R. *Axiomatics and Dogmatics*. Oxford: Oxford University Press, 1982.

Carter, Jimmy. *Always a Reckoning*. New York: Times Books, 1995.

Catton, Christopher. *The Geomancie of Christopher Catton*. Translated by Francis Sparry. London, 1591.

Cavendish, Margaret. *Poems and Fancies*. 2nd ed. London: J. Martin and J. Allestrye, 1653.

Chevreul, Michel. *The Laws of Contrast of Colour*. Translated by John Spanton. London: G. Routledge, 1857.

Cicero. *On Divination*. Translated by C. D. Young. London, 1868.

———. *On the Nature of Gods*. Translated by Thomas Franklin. London: William Pickering, 1829.

Collins, Billy. *Questions about Angels*. Pittsburgh: University of Pittsburgh Press, 1999.

Coover, Robert. "The Babysitter." In *The Norton Anthology of Short Fiction*. 2nd ed. Edited by R. V. Cassill. New York: W. W. Norton, 1981.

Copernicus. *On the Revolutions*. Translated by E. Rosen. London: Macmillan, 1972.

Cornford, F. M. *Plato's Cosmology*. London: Routledge and Kegan Paul, 1966.

Cosmas Indicopleustes. *The Christian Topography*. Translated by J. W. McCrindle. New York: Hakluyt Society, 1897.

Creeley, Robert. "Four." From "Numbers: For Robert Indiana." In *Pieces*. New York: Charles Scribner's Sons, 1969.

Dacier, André. *Life of Pythagoras, with his Symbols and Golden Verses. Together with the Life of Hierocles, and His Commentaries upon the Verses*. Translated by N. Rowe. London: Jacob Tonson, 1707.

Dales, H. G., and G. Oliveri, eds. *Truth in Mathematics*. Oxford: Clarendon Press, 1998.

Dante Alighieri. *The Divine Comedy*. Translated by Henry F. Cary. New York: T. Y. Crowell, 1897.

Davidson, Herbert A. *Proofs for Eternity, Creation, and the Existence of God in Medieval Islamic and Jewish Philosophy*. Oxford: Oxford University Press, 1987.

Debaene, Stanislas. *The Number Sense: How the Mind Creates Mathematics*. Oxford: Oxford University Press, 1997.

Diogenes Laertius. *The Lives, Opinions, and Remarkable Sayings of the Most Famous Ancient Philosophers, Made English by Several Hands*. 2 vols. London, 1696.

Donne, John. *Poems of John Donne*. Edited by E. K. Chambers. New York: Charles Scribner's Sons, 1896.

Dostoyevsky, Fyodor. *The Brothers Karamazov*. Translated by Constance Garnett. Edited by Ralph Mathew. New York: W. W. Norton, Norton Critical Edition, 1976.

Duchamp, Marcel. *Salt Seller: The Writings of Marcel Duchamp*. Edited by Michel Sanouillet and Elmer Peterson. Oxford: Oxford University Press, 1973.

Duhem, Pierre. *Medieval Cosmology: Theories of Infinity, Place, Time, Void, and the Plurality of Worlds*. Translated by Roger Ariew. Chicago: University of Chicago Press, 1985.

Edgerton, Samuel. *The Renaissance Rediscovery of Linear Perspective*. New York: Basic Books, 1975.

Eldredge, Laurence. "Late Medieval Discussions of the Continuum and the Point of the Middle English Patience." *Vivarium* 17, no. 2 (1979): 90–115.

Eliot, T. S. *The Four Quartets*. San Diego: Harcourt Brace, 1971.

Empedocles. *The Poem of Empedocles*. Text and translation by Brad Inwood. Toronto: University of Toronto Press, 2001.

Engels, Donald. "The Length of Eratosthenes' Stade." *American Journal of Philology* 106, no. 3 (Autumn 1985): 298–311.

Escher, M. C. *Escher on Escher: Exploring the Infinite*. New York: Abrams, 1986.

Euclid. *The Thirteen Books of Euclid's Elements*. Vol. 1. Translated with introduction and commentary by Sir Thomas. L. Heath. New York: Dover Publications, 1956.

Eves, Howard. *An Introduction to the History of Mathematics*. 6th ed. Orlando, Fla.: Harcourt Brace Jovanovich, 1990.

Feder, Lillian. *Ancient Myth in Modern Poetry*. Princeton: Princeton University Press, 1971.

Gafori, Franchino. *Theorica Musice*. Milan: Ioannes Petrus de Lomatio, 1492.

Galileo Galilei. *Dialogue Concerning the Two Chief World Systems*. Translated by Stillman Drake. Berkeley: University of California Press, 1962.

Ginsberg, Allen. *Kaddish and Other Poems, 1958–1960*. 4th ed. San Francisco: City Lights Bookstore, 1965.

Goethe, Johann Wolfgang von. *Conversations of Goethe with Eckermann and Soret*. Translated by John Oxenford. London: George Bell and Sons, 1875.

Grant, Edward. *God and Reason in the Middle Ages*. Cambridge: Cambridge University Press, 2001.

———. *Planets, Stars, and Orbs: The Medieval Cosmos, 1200–1687*. Cambridge: Cambridge University Press, 1994.

———. *Science and Religion, 400 B.C. to A.D. 1550: From Aristotle to Copernicus*. Baltimore: Johns Hopkins University Press, 2004.

———, ed. and trans. *Nicole Oresme and the Kinematics of Circular Motion*. Madison: University of Wisconsin Press, 1971.

Greene, Brian. *The Elegant Universe*. New York: Vintage, 1999.

Hart, Ivor. *The World of Leonardo Da Vinci: Man of Science, Engineer, and Dreamer of Flight*. New York: Viking, 1961.

Heisenberg, Werner. *Philosophic Problems of Nuclear Science*. New York: Fawcett World Library, 1966.

Henderson, Linda. *The Fourth Dimension and Non-Euclidean Geometry in Modern Art*. Princeton: Princeton University Press, 1983.

Heninger, S. K. *Touches of Sweet Harmony: Pythagorean Cosmology and Renaissance Poetics*. San Marino, Calif.: Huntington Library Press, 1974.

Heraclitus. *Fragments: The Collected Wisdom of Heraclitus*. Translated by Brooks Haxton. New York: Viking, 2001.

Hobson, E. W. *Squaring the Circle: A History of the Problem*. Cambridge: Cambridge University Press, 1913.

Huneker, James. *Pathos of Distance*. New York: Charles Scribner's Sons, 1913.

Huxley, Thomas. *Collected Essays*. 9 vols. London, 1894.

Iamblichus. "Protrepticae Orationes ad Philosophiam." In *The History of Philosophy: Containing the Lives, Opinions, Actions, and Discourses of the Philosophers of Every Sect*. 3rd ed. Thomas Stanley. London, 1701.

Jonas, Leah. *The Divine Science: The Aesthetic of Some Representative Seventeenth-Century English Poetry*. New York: Columbia University Press, 1940.

Jonson, Ben. *The Works of Ben Jonson*. 7 vols. Edited by H. R. Whalley. London, 1756.

Joosten, Joop. "Painting and Sculpture in the Context of De Stijl." In *De Stijl, 1917–21: Visions of Utopia*. Edited by Mildred Friedman. Oxford: Phaidon, 1982.

Kant, Immanuel. *Critique of Judgment*. Translated by James Creed Meredith. Web edition published by ebooks@Adelaide. http://ebooks.adelaide.edu.au/k/kant/immanuel/k16j/.

Katz, Victor. *A History of Mathematics: An Introduction*. 2nd ed. Reading, Mass.: Addison-Wesley, 1998.

Keats, John. *The Letters of John Keats, 1814–1821*. 2 vols. Edited by Hyder Edward Rollins. Cambridge: Harvard University Press, 1958.

Kepler, Johannes. *Harmonices mundi libri V*. Linz, 1619.

———. *The Secret of the Universe*. Translated by A. M. Duncan. New York: Abaris Books, 1981.

Kirk, G. S., and J. E. Raven. *The Presocratic Philosophers*. Cambridge: Cambridge University Press, 1962.

Kline, Morris. *Mathematical Thought from Ancient to Modern Times*. 3 vols. Oxford: Oxford University Press 1972.

Koestler, Arthur. *The Sleepwalkers*. New York: Grosset and Dunlap, 1959.

Kretzmann, Norman, ed. *Infinity and Continuity in Ancient and Medieval Thought*. Ithaca: Cornell University Press, 1982.

Lactantius. *The Divine Institutes.* In *The Ante-Nicene Fathers: Translations of the Writings of the Fathers down to A.D. 325.* Vol. 7. Edited by Alexander Roberts and James Donaldson. New York: Christian Literature, 1890.

Laertius, Diogenes. *The Lives, Opinions, and Remarkable Sayings of the Most Famous Ancient Philosophers, Made English by Several Hands.* 2 vols. London, 1696.

Langer, Susanne. *Feeling and Form: A Theory of Art.* New York: Charles Scribner's Sons, 1953.

Livio, Mario. *The Golden Ratio: The Story of Phi, the World's Most Astonishing Number.* New York: Broadway Books, 2002.

Lohr, Charles. H. "The Medieval Interpretation of Aristotle." In *The Cambridge History of Later Medieval Philosophy: From the Rediscovery of Aristotle to the Disintegration of Scholasticism, 1100–1600.* Edited by Norman Kretzmann, Anthony Kenny, and Jan Pinborg. Cambridge: Cambridge University Press, 1982.

Long, A. A., ed. *The Cambridge Companion to Early Greek Philosophy.* Cambridge: Cambridge University Press, 1999.

Longo, Oddone. "Ancient Moons." *Earth, Moon, and Planets* 85–86, no. 0 (1999): 237–43.

Loran, Erle. *Cézanne's Composition.* Berkeley: University of California Press, 1943.

Lovejoy, Arthur O. *The Great Chain of Being: A Study of the History of an Idea.* Cambridge: Harvard University Press, 1936.

Lucretius. *The Nature of the Universe.* Translated by Ronald Latham. Baltimore: Penguin Books, 1963.

Lynton, Norbert. *The Story of Modern Art.* Ithaca: Cornell University Press, 1980.

MacDonald, Scott. *Being and Goodness: The Concept of the Good in Metaphysics and Philosophical Theology.* Ithaca: Cornell University Press, 1991.

Macrobius. *Commentary on the Dream of Scipio.* Translated by William H. Stahl. New York: Columbia University Press, 1990.

Mahoney, Michael S. "Mathematics." In *Science in the Middle Ages.* Edited by David Lindberg. Chicago: University of Chicago Press, 1978.

Maimonides, Moses. *The Guide for the Perplexed.* 2nd rev. ed. Translated by M. Friedlander. London: Routledge and Kegan Paul, 1904.

Malevich, Kazimir. *The Non-Objective World.* Translated by Howard Dearstyne. Chicago: Paul Theobold, 1959.

Mandelbrot, Benoit B. *The Fractal Geometry of Nature.* New York: W. H. Freeman, 1982.

Mann, W. "The Ontological Presuppositions of the Ontological Argument." *Review of Metaphysics* 26 (1972–73): 260–77.

Marvell, Andrew. *The Poetical Works of Andrew Marvell*. London: Alexander Murray, 1870.

Milton, John. *Paradise Lost*. In *The Poetical Works of John Milton: With a Life of the Author ... by Charles Dexter Cleveland*. New York: A. S. Barnes, 1873.

Misner, C. W., and J. A. Wheeler, "Classical Physics as Geometry: Gravitation, Electromagnetism, Unquantized Charge, and Mass as Properties of Curved Empty Space." *Annals of Physics* 2, no. 6 (1957): 525–603.

Moore, A. W. *The Infinite*. London: Routledge, 1990.

Most. Glenn W. "The Poetics of Early Greek Philosophy." In *The Cambridge Companion to Early Greek Philosophy*. Edited by A. A. Long. Cambridge: Cambridge University Press, 1999.

Murdoch, John. "Infinity and Continuity." In *The Cambridge History of Later Medieval Philosophy: From the Rediscovery of Aristotle to the Disintegration of Scholasticism, 1100–1600*. Edited by Norman Kretzmann, Anthony Kenny, and Jan Pinborg. Cambridge: Cambridge University Press, 1982.

Murdoch, John, and Edward Synan. "Two Questions on the Continuum: Walter Chatton (?), O.F.M., and Adam Wodeham, O.F.M." *Franciscan Studies* 26 (1965): 212–88.

Nelson, Edward. "Syntax and Semantics." www.math.princeton.edu/~nelson/papers/s.pdf.

Neruda, Pablo. "Tonight I Can Write." In *Selected Poems*. Translated by Nathaniel Tarn, Anthony Kerrigan, W. S. Merwin, and Alastair Reid. Edited by Nathaniel Tarn. London: Jonathan Cape, 1970.

Newman, James R., ed. *The World of Mathematics*. 4 vols. New York: Simon and Schuster, 1956.

Nicolson, Marjorie Hope. *Mountain Gloom and Mountain Glory: The Development of Aesthetics of the Infinite*. New York: Norton, 1959.

———. *Newton Demands the Muse: Newton's "Optiks" and the Eighteenth-Century Poets*. Princeton: Princeton University Press, 1946.

North, J. D. *Chaucer's Universe*. Oxford: Clarendon, 1988.

Overy, Paul. *De Stijl*. New York: Thames and Hudson, 1991.

Ovid. *Metamorphoses*. Translated by Rolfe Humphries. Bloomington: Indiana University Press, 1955.

Peacock, Thomas Love. "The Four Ages of Poetry." In *The Works of Thomas Love Peacock*. Edited by Henry Cole. London: Richard Bentley and Sons, 1875.

Peitgen, H.-O., H. Jürgens, and D. Saupe. *Fractals for the Classroom*. New York: Springer, 1992.

Philip, J. A. *Pythagoras and Early Pythagoreanism*. Toronto: University of Toronto Press, 1966.

Piero della Francesca. *De prospectiva pingendi.* Edited by G. Nicco-Fasola. Florence: G. C. Sansoni, 1942.

Plato. *The Republic.* Translated by Tom Griffith. Cambridge: Cambridge University Press, 2000.

———. *Timaeus and Critias.* Translated by D. Lee. London: Penguin Books, 1977.

Plutarch. *Plutarch's Morals. Translated from the Greek by Several Hands.* Corrected and revised by William W. Goodwin, with an introduction by Ralph Waldo Emerson. Boston: Little, Brown, 1878.

Pound, Ezra. *Section: Rock-Drill, 85–95 de los cantares.* New York: New Directions, 1956.

Quine, Willard van Orman. "On What There Is." In *From a Logical Point of View.* New York: Harper Torchbooks, 1963.

Read, Herbert. "Human Art and Inhuman Nature." In *Philosophy of Modern Art: Collected Essays.* Freeport, N.Y.: Books for Libraries Press, 1971.

Robinson, James Harvey, ed. *Readings in European History.* 2 vols. Boston: Ginn, 1904–6.

Rucker, Rudy. *Infinity and the Mind: The Science and Philosophy of the Infinite.* Princeton: Princeton Science Library, 1995.

Russell, Bertrand. *A History of Western Philosophy.* New York: Simon and Schuster, 1945.

———. *Logic and Knowledge: Essays, 1901–1950.* Edited by R. H. Marsh. New York: G. P. Putnam's Sons, 1968.

———. "The Philosophy of Logical Atomism." *Monist* 28 (1918): 495–526.

———. "Recent Work on the Principles of Mathematics." *International Monthly* 4 (1901): 83–101.

Russell, Jeffrey Burton. *Inventing the Flat Earth: Columbus and Modern Historians.* Westport, Conn.: Praeger Publishers, 1991.

Salluste du Bartas, Guillaume. *His Devine Weekes and Workes.* In *The Complete Works.* Translated by Joshua Sylvester. Edited by Alexander Grosart. 1880. Reprint, Hildesheim: Georg Olms Verlagsbuchhandlung, 1969.

Sartre, Jean-Paul. *The Emotions: Outline of a Theory.* Translated by Bernard Frechtman. New York: Citadel Press, 1989.

Sedley, David. "Parmenides and Melissus." In *The Cambridge Companion to Early Greek Philosophy.* Edited by A. A. Long. Cambridge: Cambridge University Press, 1999.

The Seven Ecumenical Councils of the Undivided Church. Vol. 14 in *Nicene and Post-Nicene Fathers.* 2nd ser. Edited by P. Schaff and H. Wace. Translated by H. R. Percival. Grand Rapids: W. B. Eerdmans, 1955.

Shlain, Leonard. *Art and Physics: Parallel Visions in Space, Time, and Light.* New York: Morrow, 1991.

Sidney, Phillip. *Defense of Poesy: Otherwise Known as an Apology for Poetry.* Edited by Albert S. Cook. Boston: Ginn, 1890.

Simplicius. *On Aristotle's* On the Heavens *2.10–14.* Translated by Ian Mueller. Ithaca: Cornell University Press, 2005.

Spengler, Oswald. *Perspectives of World-History.* Vol. 2 of *The Decline of the West.* Translated by Charles Francis Atkinson. London: George Allen and Unwin, 1928.

Spenser, Edmund. "An Hymn in Honour of Love." In *The Poetical Works of Edmund Spenser, with Memoir and Critical Dissertations by the Rev. George Gilfillan.* Vol. 5. Edinburgh: J. Nichol, 1859.

Stanley, Thomas. *The History of Philosophy: Containing the Lives, Opinions, Actions, and Discourses of the Philosophers of Every Sect.* 3rd ed. Thomas Stanley. London, 1701.

Stein, Gertrude. "Picasso." *Camera Work* (August 1912).

Sterne, Laurence. *The Works of Laurence Sterne in Four Volumes.* Edited by James P. Browne. London: Bickers and Sons, 1873.

Stevin, Simon. "Disme: The Art of Tenths." Translated by Robert Norton. London, 1608.

Sylla, Edith Dudley. "God, Indivisibles, and Logic in the Later Middle Ages: Adam Wodeham's Response to Henry of Harclay." *Medieval Philosophy and Theology* 7 (1998): 69–87.

Synan, Edward A. "Introduction: Albertus Magnus and the Sciences." In *Albertus Magnus and the Sciences: Commemorative Essays.* Edited by James A. Weisheipl. Toronto: Pontification Institute of Medieval Studies, 1980.

Taylor, C. C. W. *The Atomists: Leucippus and Democritus. Fragments: A Text and Translation.* Toronto: University of Toronto Press, 1999.

Taylor, R., A. Micolich, and D. Jonas. "Fractal Expressionism." *Physics World* 12, no. 10 (1999): 25–28.

Tennyson, Alfred. *The Holy Grail and Other Poems.* Boston: Fields, Osgood, 1870.

Thorndike, Lynn. *University Records and Life in the Middle Ages.* New York: Columbia University Press, 1944.

Van Doesburg, Theo. "Painting: From Composition to Counter-Composition." In *De Stijl.* Compiled by Hans L. C. Jaffé. New York: Harry N. Abrams, [1970].

Van Helden, Albert. *Measuring the Universe: Cosmic Dimensions from Aristarchus to Halley.* Chicago: University of Chicago Press, 1985.

Vantongerloo, Georges. *Paintings, Sculptures, Reflections.* New York: Wittenborn, Schultz, 1948.

Veltman, Kim. *Linear Perspective and the Visual Dimensions of Science and Art.* Munich: Deutscher Kunstverlag, 1986.

Verma, Rajendra. *Time and Poetry in Eliot's Four Quartets.* Atlantic Highlands, N.J.: Humanities Press, 1979.

Vitruvius. *The Ten Books on Architecture*. Translated by Morris Hicky Morgan. Cambridge: Harvard University Press, 1914.

Wachtel, Albert. "Goddess." *Gettysburg Review* 18, no. 1 (Spring 2005): 19–31.

Weber, Maxwell. "The Fourth Dimension for a Plastic Point of View." *Camera Work* no. 31 (July 1910).

White, Andrew Dickson. *A History of the Warfare of Science with Theology in Christendom*. New York: D. Appleton, 1898.

Wilson, Robin. *Four Colors Suffice: How the Map Problem Was Solved*. Princeton: Princeton University Press, 2002.

Wren, Christopher. "Appendix: Of Architecture; and Observations on Antique Temples, etc." In *Parentalia; or, Memoirs of the Family of Wrens*. London, 1750. Reprint, Farnborough, Eng.: Gregg Press, 1965.

INDEX